Simplify

Layman's Terms Biometrics

"A Journey into Biometrics, Access Control and Time and Attendance"

NICOLAS GARCIA

1. TABLE OF CONTENT

1.	TABLE OF CONTENT	2
2.	INTRODUCTION	3
3.	UNDERSTANDING BIOMETRICS	6
4.	ESSENTIAL CONCEPTS AND LIMITATIONS OF BIOMETRICS	18
5.	PRACTICAL USE OF BIOMETRICS	31
6.	ACCESS CONTROL	36
7.	TIME AND ATTENDANCE SYSTEMS (TNA)	99
8.	LOGICAL ACCESS CONTROL / IDENTITY MANAGEMENT	107
7.	PRIVACY AND DATA PROTECTION	116
8.	FREQUENTLY ASKED QUESTIONS	120
9.	TO GO FURTHER	126
10.	CONCLUSION	136
11.	REFERENCES & CREDITS	139
12.	DEFINITIONS	141
13.	ABOUT THE AUTHOR	151

2. INTRODUCTION

Identity as far as we can remember has always been at the centre of society. The concept has been studied and debated for millennia, but even in the most so-called primitive pack where the main identity was the pack, each member had its own character by which one could relate to each other within a specific group, i.e. the fastest one, the strongest one, the oldest one, etc.
As societies and social orders grew bigger, so did the concept of **identity and population management.** This concept at the time, introduced advanced legal notions of inheritance, ownership, belonging, duties and rights, etc.
When relating to each other by use of superlatives (Conan the best, Ivan the strongest, etc.) became too limiting and impractical to continue growing, humanity had no choice but to adapt its method of identification and registration of each member of the population. With the availability of writing, the Clergy whose members were usually the only ones around to master reading and writing created the first registries. The holding of births and deaths registries in their circumscriptions was the beginning of **population** and **identity management**.
Identification documents were subsequently introduced, and secret organisations started using secret codes to identify its members and ensure that they were allowed where their meetings were taking place. **Access Control was born**.
Furthermore, as the economy started to grow exponentially, workers were identified through a registry to ensure their labour was accounted for and remuneration was paid to the right person. This was now the birth of **Time and Attendance**.
Over the past few decades, safety & security have become increasingly important worldwide and have pushed traditional methods to early retirement. More advanced requirements and standards have evolved, which are increasing the cost and complexity of compliance.
New systems have emerged, combining strong benefits of traditional principles with innovative technology and science, bringing more added value and benefits to the core functions of **Access Control** and **Time and Attendance**.

The added benefits can range from assisting businesses ensuring the safety and security of their employees and visitors to protecting their confidential data, assist in complying with legal & regulatory requirements in their respective countries.

Biometrics technology adoption has grown at a phenomenal rate over the past few decades, and it seems unlikely that this trend will decline any time soon.

An entire market has evolved around biometric access control and time & attendance. This book aims at giving you an overview of the technology, how it works, and how you can take advantage of it to improve your business.

Why this book?

Over the years, I have interacted with thousands of people through meetings, conferences, training and many other social events. In many cases people did not understand the concept of biometrics but wanted to learn its practicalities. Many people also thought they knew everything about biometrics but were often falling short of really understanding how biometrics could help them and what its limitations were. After over fifteen years of Biometrics exposure, I felt that it would be a good idea to write a book and impart my knowledge and experience with many more people than those that I was blessed enough to meet in the past.

Is this book for you?

I wrote this book with two things in mind: "Practicality and simplicity".
This book is for those who want to understand what Biometric Technologies are about and, how they can be of benefits in the context of Access Control and Time and Attendance.
This book is not a scientific book; therefore, you will not find anything complicated such as formulas to calculate the speed of a system. I tried to keep it simple.
This book is product and modality agnostic and does not refer to any specific manufacturer or describe my preferences in term of brands, but it should make you better equipped to determine which product is better suited for your requirements.

3. UNDERSTANDING BIOMETRICS

What is "Biometrics"?

Let's break the word "Biometrics" down and try to understand what it entails. Biometrics is the combination of two words:

- Biology which stands for "Life".
- Metrics which stands for "Measurement".

By extension, "Biometrics" means the measurement of life.
In other words, "Biometrics" is a method used to measure the physical or behavioural characteristics of an individual.
To be considered a Biometric feature, at least two characteristics must be present:

1. The feature should prove to be unique enough that no two individuals share the same, or that the uniqueness of the biometric is quantifiable.

2. The feature should be immutable and not fundamentally change during its lifetime.

If the characteristics are unique enough, biometrics can then be used to identify or verify the identity of an individual amongst a predefined population in a repeatable and consistent way.

"Population" here widely refers to a group of people that can be:

- A country's population,
- Members of a club or organisation
- A particular group of individuals such as a forensic / police criminal database,
- The residents of a Residential Estate,
- The employees of a company,
- The customers of financial institutions,
- A collection of individuals.
- A population can, therefore, range from a few individuals to billions.

Biometric technologies are used in computer science as a form of identification and access control. It is also used to identify individuals in groups that are on watch-lists (terrorists, criminals, sex offenders etc) and keep the general public safe.
For convenience purpose, in this book and unless otherwise specified, we will refer to fingerprints or a fingerprint-based system. However, the principles depicted are very similar irrespective of which biometric features are used.

Why Biometrics?

Biometric technologies are crucial to combining security & convenience but also and most importantly, Biometrics is the only way of ensuring the unique identity of an individual.
Nowadays, Biometrics technologies are increasingly becoming essential in the business world to ensure proper access control, employee Time and Attendance and or Time Management solutions.
In everyday life, Biometrics is growing very popular and embedded fingerprint readers and facial or iris recognition technology now regularly feature on cellular phones.

Are there different types of biometric?

Biometric features fall into two categories: "Physical and Behavioural"

Physical (What you are)

- Fingerprint
- Vein
- Facial
- Iris
- Voice
- DNA
- Etc.

Behavioural (How you behave / How you do something)

- Gait
- Signature
- KeyStroke
- Etc.

These different features are also commonly referred to as modalities.

Physical feature biometrics tend to be much more accurate and reliable than behavioural biometrics.
There is no one size fits all when it comes to applications requiring biometric so, you will find a list of noteworthy biometrics with their pros and cons later in this book. This will help you find the best biometric for your requirements.

How do Biometric systems work?

Modern biometric systems are designed to provide fast and reliable tools to authenticate[1] one individual within a population. It is performed through identification or verification of a claimed identity.
To that effect, the system should be able to consistently and repeatably transform analogic or "natural" biometric features into computer-readable data.
Biometric systems function on the same model through three distinct phases:

- CAPTURE
- ENCODING
- MATCHING

Each phase and how they interact with each other will now be covered in more detail.
For this explanation, we will generally talk about optical fingerprint biometric system.
The principle of the process can, however, be adapted easily for other biometric types.
Some naming might be specific to each biometric type.
All three components will also be covered separately for ease of understanding. It is not uncommon however to find those components physically located and integrated into the same device, i.e. on the same computer or access control reader.

---

1 - To Authenticate is the action of ensuring the veracity of something or someone's identity

Did you know?

Computers can only interpret electrical current. They recognise if a switch is "on" or "off".
It is called the binary mode and represented by "1" and "0". Each iteration is called a bit.
This concept is essential to understand because the whole automated system stands on the ability to convert a biometric feature in a computer-readable format.
Computers see zero as "the switch is off", and one as "the switch is on".
The next major breakthrough will be "Quantum Computing". Quantum Computing goes beyond the current limitation and allows a bit to be "1 & 0" at the same time. This will be the next major leap forward in processing power.

Capturing

The "Capture", "Acquisition" or "Extraction" of the biometric feature is usually performed during the "Enrolment" phase. Enrolment is the process of registering an individual into a system. The data captured typically includes : standard demographic information such as name, surname, age, gender, as well as some unique identifier such as an employee number, and ultimately the biometric feature itself.

At this stage, the system only acquires data and digitise the biometric feature into a generated raw file such as a picture of the biometric feature. The part of the system managing this feature is called an "imager" when it acquires a picture or an image.

In the case of a fingerprint, the picture would be a digital file (BMP, JPEG,...) made of "0" and "1", e.g. "1010110011" or in other words, a format that computerised system can understand.

This part of the process is critical as without a quality acquisition, the rest of the system cannot function optimally. As the saying goes: "Garbage IN, Garbage OUT".

In a fingerprint-based system, the size of the sensor can sometimes be critical to ensure the capture of a maximum of information during enrolment, especially in rough environments where fingerprints can be damaged.

The illustrations below show a potential loss of crucial information when enrolling with a small sensor.

Small sensor Capture

Large sensor capture

Encoding

After image acquisition, a component called "encoder" or "feature extractor and encoder" performs the next function.
During the encoding phase, unique biometric features are detected, extracted and transformed into another digital format called a template. This template can be of different formats optimised according to the intended use of the template.

  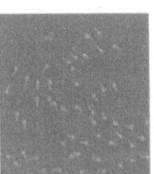 

1 - Acquisition & Minutiae Detection   2 - Image Destruction, Minutia Matching and Comparison   3 – Final Encoded Template   4 – Binary File Coded Template

The encoder runs through different processes of cleaning, healing and converting the image. The "recipe" followed during these processes is called an "algorithm".
The encoder's ultimate purpose is to take a raw image and to generate a template amongst many such as:
The manufacturer proprietary format (PKCOMP, ...)

- ISO format
- ANSI format
- WSQ format
- Etc

The manufacturer proprietary format is usually the most performant of all available formats because it is optimised to perform on the manufacturer hardware. However, for compatibility concerns, it is sometimes necessary to use more standardised formats.

The encoder should be capable of generating multiple formats during the same process if necessary, and the raw image should be discarded or preserved, according to the end user's requirements and local regulations.

Like the imager, a great encoder can make the difference between a suitable system and a bad system.

Indeed, the encoder oversees the enhancement of images captured by the imager when necessary.

Although it is always better to have both a good imager and a good encoder, a great encoder can help mitigate an average imager.

After encoding is complete, the original image can be kept or deleted, depending on the application.

Standard practice is to only keep the original image for civil and forensic application purposes.

Once a template is generated, it cannot be reversed engineered into the original source image, or in other words; the original image cannot be recreated exactly from the template.

Did you know?

Sometimes people swap the terms "template" and "algorithm". This is an abuse of language as an algorithm is a process and set of formulae, used to generate a template. An algorithm can also be compared to a cooking recipe. It is all the steps that the system must follow and all the resources to be used to get to the result.

The template is then stored in a database or any number of media for future reference.

Examples of template storage options include (non-exhaustive):

- On a Database (central or distributed)
- On a terminal (kiosk, time clock, access reader etc.)

- On a smart card / token (access card, key fob, credit card etc.)
- On a barcode (such as an airline boarding pass)
- On a mobile device (cell phone or tablet)
- In the cloud

The size of the template will depend on the type of biometrics captured as well as the level of compression of the file. It can range from a few bytes (128bytes for instance) to several Kbytes (64Kbytes or more for instance).

Matching

Once the biometric features have been captured and encoded into a template, the system uses another component called the "Matcher".
Simply put, the matcher manages the search and finds a match ("HIT") or a no match ("NO HIT") between the biometrics presented to the reader and the templates encoded before that.

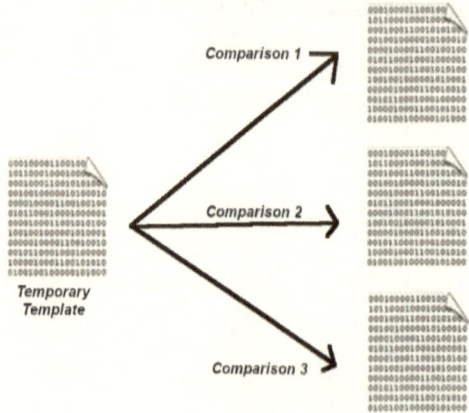

It is essential to understand that many parameters can influence the quality, performance, and accuracy of the matcher's results.
In an ideal world, the imaging/encoding/matching trio should always work 100% accurately.
However, it is not always the case in a real-world environment, and there are many good reasons why.
Two different methods are available for matching:

1. Verification (1:1)
2. Identification (1:n or 1-to-many)

Verification or 1:1 (pronounced one-to-one)

Verification is the process of checking the identity that someone claims to hold.

For example, when a traveller presents his passport at an immigration counter, he claims to be the passport holder and the immigration officer performs a visual verification by checking what is inside the travelling document and who is in front of him.

If the verification is positive, the traveller can move on. If not, other procedures kick in.

To perform a 1:1 Verification, the system will first have to proceed following a similar process to the one used during the enrolment phase.

However, the system will only store the template temporarily in a memory buffer zone and will discard it on completion of the operation. The process usually takes place within the device RAM (Random Access Memory).

The system will then compare the original template with the one that has just been created and stored in the buffer zone and eventually come up with an answer: "MATCH/HIT" or "NO MATCH / NO HIT".

To perform a 1:1 Verification, the user needs to first provide a token for the system to know which fingerprint to look for in the database.

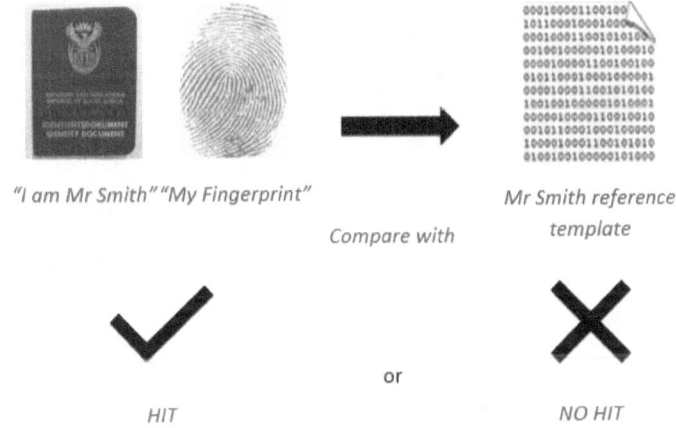

"I am Mr Smith" "My Fingerprint"   Mr Smith reference template

Compare with

HIT   or   NO HIT

Identification or 1:n (pronounced "one-to-n" or "one-to-many")

"Identification" is the process of finding an individual's identity in a database of individuals credentials without having any prior indication of who the person is.

A "1: n Identification" process is a succession of "1:1" verifications, one after the other until a result is found (called a "MATCH") or until the end of the database has been reached with no success ("NO MATCH").

"HIT OR NO HIT" can sometimes be used in instead of "MATCH or NO MATCH". The main difference is that no one needs to present a token[2] to perform an identification.

Although it seems as simple as verification on paper, the process is undoubtedly more resource intensive and more complicated than verification. The difficulty in retaining performance and reliability increases exponentially and proportionally to the database size. That is why, although it is relatively easy for a manufacturer to propose identification on a dozen of individuals, not everybody can propose a reliable and fast solution for hundreds of thousands, millions or even billions of individuals, depending on the application requirement.

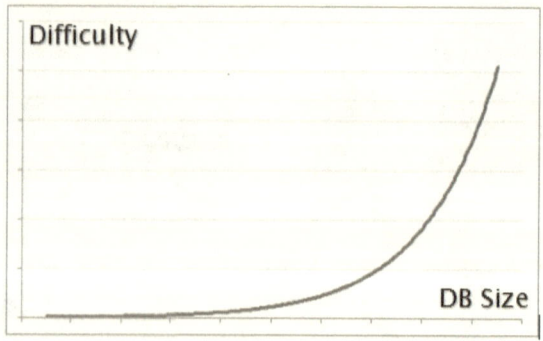

Did you know?

---

2 A Token can be a card, tag, ID number, or anything similar

We call "authentication" the process of verifying the identity of somebody. Verification and Identification are both ways of authenticating somebody. Authentication is also used to ensure that electronic transaction or documents are genuine.

4. ESSENTIAL CONCEPTS AND LIMITATIONS OF BIOMETRICS

Contrary to public belief, Biometric Systems by their very nature cannot be perfect and this for several reasons:

1. The system only matches what is already in a database.
2. The system can only be as good as the quality of data captured into the system.
3. Similarly, the quality and amount of data can change from one capture to another, depending on external factors such as dust, light, injuries, sickness, and so forth.

It, therefore, follows that each biometric technology has its advantages and disadvantages, based on the requirement of every user.
We will now cover the reason why a system cannot be perfect more in details.

Biometrics in a REAL-WORLD environment

Biometric systems are subject to real-life influences, and therefore performance can be directly affected by these elements. Earlier in this book, we described the principal of biometrics and we will now challenge them based on elements that can affect the system.
These examples are not exhaustive, but if one understands how they affect their specific biometrics features, one should be able to deduce other issues that could be faced.

Fingerprint imperfection

The quality of a fingerprint can vary based on external elements such as:

- Cleanliness of the finger
- Injuries
- Chemicals
- Harsh work environment
- Other environmental factors such as light and dust
- Disease
- Age
- Etc.

As biometric systems read fingerprints across different times and places, the generated binary file will be affected by such conditions. As illustrated below, although the fingers are the same, generated templates are different. Furthermore, when a finger is placed at a different angle, the grid used to record minutiae points will not be the same and therefore the binary file could be different for each new scan.
Based on these observations, how can biometric systems match two binaries which are so different from one another?

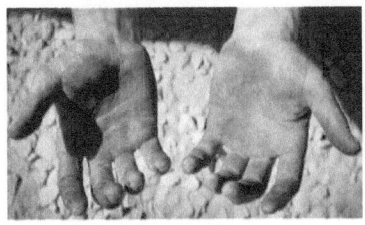
Dirty hands

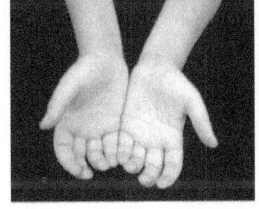
Clean hands

Dirty Finger	0	1	1	0	1	
Clean Finger	1	0	0	1	0	1

In the above-simplified example, both fingers are the same but because of the dirt, the binary might be slightly different.

To allow the system to function, it is, therefore, necessary to define an acceptable level of variance between the different templates and introduce a notion of Security vs Convenience trade-off.

Security vs Convenience

A computer system can only make a decision based on pre-defined parameters.
Depending on the use case, a trade-off between security and convenience is essential to run the system optimally, and the computer can only decide if it knows where to place the cursor.
On a scale from 0 to 100, convenience might be at a maximum of 30% importance, and security should logically stand at 70% minimum.
It is essential to understand the correlation between the two parameters. One is inversely proportional to the other, and therefore an extremely secure system cannot be extremely convenient at the same time:
The more the convenience, the less the security

Further concepts with biometrics

Statistics in Biometrics

Setting a tolerated level of convenience in the system is proof if still needed that no biometric system is perfect. If it werewas perfect, the system would always be fully securitysecure while offering maximum convenience to all users.

Biometric systems rely on advanced statistical algorithms to measure the level of accuracy of a reading. The system is set to accept a result based on a pre-configured threshold (False Acceptance Rate).

In theory, this proportionally results in an opposite falsely rejected result (False Rejection). These notions of "False Acceptance" and "False Rejection" will be explained shortly.

There is no need to panic however as these "errors" are very theoretical. Based on the setting of the system it simply means that if one places a finger on the sensor successively a million times and in the same position, it might result in one of the readings being inaccurate.

It remains therefore critical to choose the right technology and the right product adapted to each use case's requirements.

Some systems allow for setting up those rates from one per cent (1%) error rate to one per 10 million. Some others only allow for one per cent to one per thousand. As with everything, performances and reliability often come with a different price tag.

In the same way, Fingerprint technologies are very fast and therefore adapted for access control purposes.

On the other hand, the error rate is much lower for instance with DNA, but DNA is not fast enough for access control. Ultimately, the type of application should dictate the choice.

False Acceptance Rate (FAR)

A direct consequence of setting this level in the system is that in some instances, two generated templates might be identified as the same, although they are not from the same original finger.
The number of errors is known as False Acceptance Rate or FAR. FAR is always shown as a statistic and should be as low as possible. For instance, if we set the system to accept a 1% FAR, it means that statistically, one of every hundred scans will wrongly identifying a finger. The higher the FAR rate, the more convenient the system as it would tolerate more difficult fingerprint.
In real life, the FAR is usually much lower than that and should remain at an acceptable level.

False Rejection Rate (FRR)

The False Rejection Rate or FRR is a direct consequence of the FAR rate and means precisely the opposite.
On a 1% FRR rate, for instance, it is admitted that for every 100 times a valid finger is presented in a sequence, one scan may be falsely rejected as not part of the system one time although it is genuinely in the system.
As for the FAR, the FRR is usually much lower than that and should always remain at an acceptable level.
FRR is the element which is most complex to quantify as it is more susceptible to external factors than FAR.
In other words, in a well-designed system, FAR is highly manageable. However as an example, the same system at a given FAR with the same readers and users may experience a different FRR if the system is deployed in a dirty factory compared to a clean laboratory.

Equal Error Rate (ERR)

The Equal Error Rate or ERR means that both the FAR and the FRR are at the same level (50% each) and that in presenting 100 times the same fingers in a row, one has as much chance to be accepted while not in the system than being rejected while genuinely being in the system.

Failure to Enrol (FTE)

Failure to Enrol or FTE refers to the rate of people that are impossible to enrol in a biometric system. There are many reasons why some people cannot be enrolled, and a few are listed below:

- Missing fingers, missing hand, missing eye.
- Somebody who is mute cannot use a voice-based biometric system.
- Tiny fingers can be hard or impossible to enrol.
- An extreme case of worn fingerprints (due to chemicals or working in rough environments)
- An extreme case of worn fingerprints due to health treatments (chemotherapy, etc.)

FTE applies to all modalities and not only to fingerprint modalities. Most reputable manufacturers will propose alternatives to cater to those difficulties.

Spoofing and Spoof Detection

Spoofing a system means to mislead the system in believing that the transaction is genuine and that the system should authenticate rather than reject it.
To spoof biometric systems, criminals do not come short of ingenuity. For instance, people use fake fingers made of different materials such as glue, latex, paper, etc.

Although creating a realistic enough fake finger to spoof biometric system is not as easy as one can see in movies, it is possible for someone who has the right hardware, materials, skill and follows the right methodology.
To protect against this risk, reputable manufacturers of biometric systems propose different types of Fake Finger Detection system referred to as FFD. Those systems can be:

- Software FFD (SFFD)
- Hardware FFD (HFFD)
- Other / Hybrid FFD

Software Fake Finger Detection (SFFD)

Software FFD is a feature in the manufacturer algorithm which attributes a score to the captured template based on common fake finger attacks previously observed and compares the score according to the level of security required. If the comparison falls between the accepted score, the process goes on, but if it falls behind, the process stops as the algorithm will consider that the fingerprint placed on the sensor is fake.
This feature is better than the standard feature but is not a silver bullet and will not recognise all fake fingers presented to the system.
SFFD is, however, an inexpensive alternative to the more effective Hardware Fake Finger detection systems which are much more precise.
In some cases, this feature is present as standard but must be activated and may lead to a (barely visible) slowdown of the system due to the increase of parameters that the firmware has to process, without any hardware upgrade. The impact may vary from one manufacturer to another.
As the security increases and therefore the convenience decreases slightly (see FFR and FAR paragraph), SFFD may increase the FRR.

Hardware Fake Finger Detection (HFFD)

Hardware Fake Finger Detection is more advanced than SFFD and incorporates extra elements such as an optronic layer which looks like an electronic circuit, or optical components (LED's & sensors).
The main difference between SFFD and HFFD is that HFFD will first check the physical properties of the presented fingers in order to establish if the finger is real or not.
Once the system concludes that the finger is real, it proceeds to enrol or authenticate an individual.
HFFD is also able to detect more Fake Finger types as this method uses real measurement as opposed to a database of previously observed results.
As the security increases and therefore the convenience decreases slightly (see FFR and FAR paragraph), HFFD as SFFD may also increase FRR.
Hardware Fake Finger Detection is the most accurate and reliable Fake Finger Detection System out there, but it comes at an extra cost.

Others or Hybrid Fake Finger Detection

As a bid to keep costs under control, manufacturers have come with different hybrid Fake Finger Detection System based both on Hardware and Software. In this instance, a slight hardware modification allows the system to differentiate the result obtained through this hardware compared to the result obtained with regular hardware.
The result then obtained is eventually compared against the database of known Fake Finger template characteristics. If the parameters are right, the system considers the fingerprint as real and goes on with the process.
An hybrid mode is not as powerful as the Hardware Fake Finger Detection system but comes at a reduced price and slightly more efficient than simple Software only Fake Finger Detection system.

Extra information about Fake Finger Detection

Although Fake Finger Detection systems have been available for years, most systems out there have been working with the standard features.
The main reason is that a biometric system is never an isolated system. It always forms part of a broader system and therefore it is often unnecessary to add the cost of the full Hardware Fake Finger Detection system to the standard system.
Most systems now tend to embed Software Fake Finger Detection by default for the peace of mind of the end-user.
The primary issue where the security breach can come from remains collusion between system administrators/operators and users.

Liveness detection

Liveness detection is a subset of Fake Finger Detection and is a feature which purpose is to ensure that not only the fingers are real but that they are part of a real individual.
For instance, with a system capable of capturing four fingers at a time, the liveness detection feature will ensure the sequence of fingers are correct (index, major, ring, pinkie) to avoid that two people try to capture part of their fingerprints together as one individual.
At the date of writing this book, many manufacturers falsely advertise FFD as being the same as liveness detection. It is however important to state that there is a difference between spoof detection and liveness detection. Liveness detection based only on sequence checking is not a real assurance of liveness. The above points are still one of the major misunderstandings in the market.

Juvenile and small fingers issues

The development of a fingerprint is comparable to the development of a flower.
As such, the fingerprints develop in the womb as the foetus is developing and continues to bloom and grow at birth.
For that reason, to enrol a baby, the heel may be preferred as it is one of the only parts of the body grown enough to be readable by the system. At birth and until about six years old, a fingerprint is too small to be enrolled. Furthermore, the fingerprint continues to grow for many years and only stop growing around adult age.
This example, however, is a rule of thumb and fingerprints of a chubby toddler with big fingers might be enrollable whereas the fingerprint of a very thin 18-year-old individual might prove challenging to enrol.

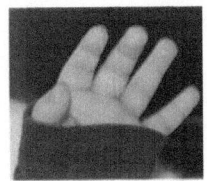 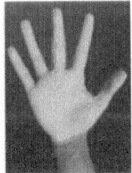

The fingerprint grows as the individual grows older, the template will change, and it might be necessary to enrol the individual again in future. It is sometimes necessary to repeat the operation several times during growth to ensure the system can still authenticate the individual efficiently and reliably. Some manufacturers have developed specific algorithms which can artificially determine what a current fingerprint will look like in the oncoming months or so and by doing that, delay or eliminate the need for further enrolment. The name of this mode is usually "Juvenile Mode".
It is available for different biometrics such as facial recognition for instance. The difference of quality between algorithms can save the system administrator from a few weeks to a few years of re-enrolment. As for the quality and reliability of FAR and FRR, these performances can sometimes come with a significant cost difference.

As the saying goes, time is money, and it sometimes costs a bit more to acquire the right system (Cost of Acquisition) but save the hassle to spend more time and money later to operate the system (Cost of Ownership). The principles of Cost of Acquisition and Cost of Ownership are covered in detail later on in this book.

Multimodal Biometrics

To fight against spoofing, FTE, FRR and FAR, manufacturers are continually releasing innovative solutions. One of them is called "Multimodality". Multimodality is the combination of multiple biometric features with the aim of mitigating the shortfall of both features while increasing performances and accuracy.
Spoofing becomes more difficult with the need to spoof two biometric features at the same time as opposed to only one in monomodal use.
Failure to Enrol rate declines; if one biometric feature fails, one can still rely on the second one.
False Rejection Rate and False Acceptance Rate are reduced as the system checks on more parameters than in monomodal system and combine performances of both.
Some examples: Fingerprint and Finger vein / Face and Iris

International Biometric Standards

What are International Standards?

In general, International Standards assure users that certified products meet or exceed minimum interoperability and quality requirement of these standards.

Different level of qualities can be experienced within the range of certified products and international standard agencies usually publish regular test results with classification and awards to the best products and solutions provider.

Why should I be interested in international standards from other countries?

Reputable country agencies such as Bureau of Standards, Federal Bureau of Investigations (FBI), and so forth, usually devise International Standards. Although these standards are not necessarily applicable or enforced in all countries, they tend to be followed and required by government departments or enterprise level companies in many other countries when they put out tenders.

Likewise, if a company wants to do business in one of the countries following those standards, it would have to be compliant as well as certified. Certifications cost money, time and effort but are necessary to ensure that products meet the adequate level of quality and performances.

US Standards

FBI PIV IQS Certification stands for "Federal Bureau of Investigation, Personal Identity Verification, and Image Quality Standard".

According to the FBI Biometric standards website, there are two standards currently encompassed by PIV IQS.

There are two standards currently in use for fingerprints: Appendix F and PIV-071006.

Appendix F has stringent image quality conditions, focusing on the human fingerprint comparison and facilitating large-scale machine many-to-many matching operation.

PIV-071006 is a lower-level standard designed to support one-to-one fingerprint verification. Certification is available for devices intended for use in the FIPS 201 PIV program.

https://www.fbibiospecs.cjis.gov/Certifications/FAQ

German Standard

The Federal Office for Information Security (German: "BundesamtfürSicherheit in der Informationstechnik", abbreviated as BSI) is the German upper-level federal agency in charge of managing computer and communication security for the German government.

The framework document TR-03121-1 gives an overview of the contents and usage of this guideline.

The current version contains the application profiles for the following areas:

Enrolment and checking of biometric data for the German Electronic Identity Card, the Electronic Passport, and the Electronic
Residence Permit
Enrolment and checking of biometric data for the EU visa information system
Enrolment of biometric data for the arrival attestation document
Enrolment and checking of biometric data in watchlist scenarios
The official page is available at the following web address:
https://www.bsi.bund.de/EN/Publications/TechnicalGuidelines/TR03121/BSITR03121.html

Indian Standard

STQC (Standardization, Testing and Quality Certification) is the Indian Ministry of Information & Communication Technology department in charge of devising standards to be adhered to when dealing with Indian Government agencies or with businesses requiring such certification.

Here are the various e-Governance applications in India using Biometrics:

- UIDAI(Aadhar)
- NPR (National Population Register)
- E-Passport
- PDS (Public Distribution System)
- RSBY (RashtriyaSwasthyaBimaYojna) Transport department for issuing or renewing Driving License
-

For more information, please visit:

http://www.egovstandards.gov.in/biometrics

This list is not exhaustive. More standards are available and published regularly.

PRACTICAL USE OF BIOMETRICS

Identity Management for commercial use

Before jumping into the different types of practical use cases where biometric technology can be useful, we will depict a few financial concepts that typically have an impact on the type of system implemented and the type of products associated with it. The role of this section is not to tell the reader which products to use but rather to create awareness of what parameters to consider before deciding on and acquiring the right solution.

Financial concepts

Return on Investment (ROI)

In a commercial environment, the notion of return on investment (ROI) is paramount.
The ROI is simply put the value recovered from an investment over a certain period in return for an investment.
When the ROI can be quantified, it becomes easy to estimate:

- How long should amortisation of investment take?
- What maximum value should be invested?

This return can be of a monetary value or tangible added value for the investment effort.
For instance, for an investment of 100USD returning 10USD back per day, it would take ten days to get the investment back and gain from the following day. Likewise, with one computer and a desk in an office, it would be unwise to invest hundreds of thousands of dollars to secure the premises.
On the other hand, if millions of dollars' worth of stock and sensitive information, it would probably be justified to invest a couple of million in a security solution.

When the ROI cannot be quantified, the following can result:

- Nothing happens
- Bad or irrational decisions tend to be more likely
- Inadequate specifications.
- Variation orders might be necessary during the project which could lead to higher costs

Cost of Acquisition

The cost of acquisition refers to all expenses accounted for during the initial procurement of a product or a solution.
It can be:

- Cost of the goods / solution / licenses
- Accessories and third-party products
- Design and selection cost (usually consulting fees and tender process)
- Installation costs
- Commissioning costs
- Maintenance costs
- Training of users, operators and if need be, administrators of the system
- Etc.

Very often, the cost of acquisition is underestimated, and extra hidden costs may arise later.
It is critical to design the system properly to ensure that the cost of acquisition is as real as possible. It is very easy for instance to forget a license or other accessories which will increase the cost at a later stage.

Cost of Ownership

The cost of ownership is the sum of all costs incurred from the moment of commissioning of the system and until the system replacement or decommissioning.
The Cost of Ownership is even more important than the cost of acquisition as it includes all the costs from the beginning to the end.
Cost of ownership should always be planned during the acquisition phase to be as cost effective as possible. This is however only possible through proper planning and requires time and discipline to avoid taking shortcuts that could prove costly later. The typical mistake is to use cheap products for cost-saving purposes but ending up by replacing the units several times during its lifespan due to poor quality or poor performances.

There are always cost implications, even when the product is under warranty:

- Call-out fees
- Labour cost
- Business disruption and contingency plan cost
- Staff loss of confidence in the system, etc.

One way to minimise the impact of call-out fees and labour charges is to sign-up an SLA (Service Level Agreement) with the service provider.
The principle of an SLA is simple. The end-user agrees to pay a negotiated amount through regular payments during a set period while the contractor agrees to render a service against these fees.
For instance, the contractor commits to visit the site twice a month for two-hours and perform preventive maintenance (if necessary) for the next two years while the customer agrees to pay the contractor a set amount of 200 USD per month during the same period.
The contractor secures an income stream, even if there is no maintenance to be done during those visits, and the customer gets the peace-of-mind that the site will always be up-and-running. SLA should be proposed at a better rate than on a case-by-case basis. Proper planning is here again the key to success.

"POOR PLANNING = POOR PERFORMANCE."

ACCESS CONTROL

Definition of Access Control

One cannot talk about "Access Control" without first defining what it is. Access Control is the means employed to control who has access to physical or logical resources by answering the questions "What, who, when, where?":

"What," can be physical (tangible) or logical (intangible) resources.

Physical resources can be assets, money, documents, etc. and logical resources can be digital files, passwords, etc.

"When," can be categorised as date, time, or ranges of each.

For instance, access every Saturday, from 8 am to 8 pm.
When can be in the **PAST** for audit purposes: "When did an event happen? ".
When can be in the **PRESENT** for current rights: "When can I access something or somewhere?"
When can be in the **FUTURE** for planning purposes: "When will he be able to access?"

"Where," is the location to be accessed.

This location can be physical (a building, a room, etc.) or logical (a computer, a hard drive, a printer, etc.)

We then speak about Physical Access Control Solutions (PACS) and Logical Access Control Solutions (LACS). By nature, Access Control always implies at least the securing one resource.

In Access Control, every "access in" or "access out" is referred to as a door or a gate. Moreover, an Access Control system can be designed to work with multiple token requirements:

- Who am I? (Biometric feature)
- What do I have? (Do I have a card or any other token?)
- What do I know? (Do I know a pin code, a password or a passphrase?)

These can be combined according to the security requirement but also according to regulations.

Components used in Access Control

Relay

A relay functions on basic electricity principles. Although it is not always visible due to miniaturisation, a circuit is a loop with three connection points. The three connectors are namely:

- Common
- Normally Open
- Normally Closed
-

Examples:

The "Common" connector is always connected, and the relay creates a loop between the common and one of the two remaining connectors "Normally Open" and "Normally Closed."
If the switch is "Normally Open," no electricity will pass through unless the relay is triggered.
If the switch is "Normally Closed," electricity will pass through until the relay is triggered.

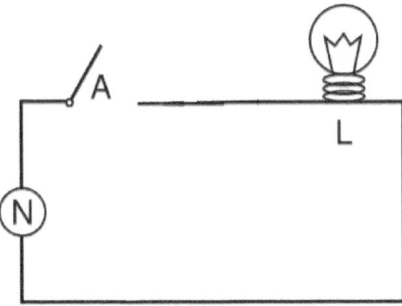

In this illustration, when the relay is open, electricity does not go through, and the light is off.
When the relay closes, electricity goes through, and the light is on.

Magnetic Lock (Maglock)

Maglock is the contraction for "magnetic lock".
The Maglock is connected in a "normally closed" mode with a relay.

The Maglock remains locked if the electric current runs through it. When the relay is triggered, it opens-up the loop and breaks the electric current flow which releases the Maglock grip and opens the door.
A Maglock strength is measured in Kilograms Force (Kgf) or Pounds Force (Lbf).

- Micro Size: 275 lbf (1,220 N) holding force.
- Mini Size: 650 lbf (2,900 N) holding force
- Midi Size: 800 lbf (3,600 N) holding force
- Standard Size: 1,200 lbf (5,300 N) holding force.
- Shear Lock: 2,000 lbf (8,900 N) holding force

The Maglock is an easy and secure solution to put in place, but it does suffer from a few shortfalls:
If there is no electricity, the Maglock will release as soon as the electromagnetic field fails.
If a Maglock rating is 300Kg and a pressure of over 300Kg is applied, the maglock will open
If the Maglock does not let go, the installation bracket could break.
If the door is not strong enough, it could break, defeating the purpose of utilising a Maglock, .e.g. with doors made of cheap wood.
A Maglock solution is as secure as the weakest link of the solution.
Maglocks are also available as a monitored maglock, equipped with a LED light that turns green when the maglock is closed and red when the maglock is opened.

Door Strike

A door strike works on the reverse principle of a Maglock.
The relay also connects in series with the door striker, but the relay state is open by default instead of closed, and as a result, the current is not going through the circuit (not energised). When the relay closes, the current runs through and releases the door striker which releases the door. From a security point of view, door strikers tend to be more secure than maglocks due to its design: Door Strike remains locked even in case of power failure
Door Strike can be interlocked quite deeply into the door, increasing the pressure needed to open the door

Door closer

A door closer is a device installed between the door and the wall to ensure that the door closes automatically when somebody opens the door and forgets to close it behind.

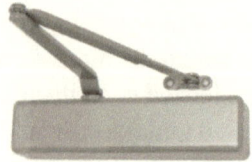

Emergency override

An emergency override is a box containing a switch behind a security window made from glass or plastic, depending on the model. In case of emergency, the glass can be broken with finger pressure or with a special mallet positioned next to the box (again depending on the model used). When the window is broken, the switch is activated, and an alarm can be triggered, and a door released for instance.Some alternatives exist where the glass can be lifted to take what is inside. The advantage is that nothing needs to be broken. The inconvenience is that the content of the box is not as secure. Sometimes the emergency box is bigger and does not hold any switch but rather emergency equipment such as a hammer to break a window or a first aid kit or any other useful items that need to be kept safe until an emergency occurs.

Emergency exit

An emergency exit is a specially marked door, conveniently placed to allow for the fast and easy exit in case of a disaster such as fire, earthquake, and so forth.

Legislations differ per country, but some particulars are widely accepted and used as a minimum standard.

For instance, this exit can only be used from the inside and must be appropriately marked as an emergency exit.

Emergency exits are fitted with what is called crash bars which release the door when pushed.

These exits are typically monitored by an access control system or door monitors, which will sound an alarm if opened.

Push-button

A push-button, as its name refers to, is a button that a user pushes to open a door. The push changes the state of a relay.
The push-button will switch the relay from **"Normally closed"** to **"Normally open"** to release a maglock and from **"Normally open"** to **"Normally closed"** to activate a door striker.
When hygiene is a concern, traditional Push-Button are falling out of fashion and are progressively replaced by so-called "no-touch" or "touch-less" buttons.

Security Remote control

A security remote control works much like a car key remote control. It has a casing containing a transmitter and one or several buttons used to send a signal to a remote device (the receiver) via radio frequency on a specific channel to a receiver.
The receiver can be used to trigger a relay, and this relay can connect to a maglock, a door striker, a motor, an alarm (as a portable panic button).
To communicate, both the transmitter and receiver must be using the same frequency (433Mhz for instance) which is usually manufacturer and model dependent. This means that **not** all remote will function will all receiver. Each button transmits on a different channel.

The remote control must be programmed in order to link it to the receiver, and that will also depend on the type of remote control used:

Switch remote control

- There are small switches on both the remote control and the receiver. They must be set up to the same positions on both sides (1=1; 2=0; 3=1; 4=0 for instance).

Self-learning remote control

- A Self-learning remote control copies the signal of the receiver automatically when both are in "learning mode." How to activate the learning mode is peculiar to each manufacturer. To know more, have a look at the manufacturer guide.

For further security, manufacturers have introduced a rolling code mechanism. Each time a remote performs self-learning, an extra security code is generated and shared between the remote and the receiver.

Did you know?

RFID signals can be jammed.
Thieves are now using ingenious methods to scramble car keys signal when you park your car.
As you walk away and push the button to lock your car, you are under the impression that your car is safe but if a jammer is used, your car will not be locked, giving easy access to your belongings.
Always double-check if the car is locked before walking away.

Key Switch box (Overriding mechanism)

A key switch box is an accessory that can be installed as part of an access control system at the entrance of a building. If the access control system fails, a special key can be used to unlock and bypass the security system.

Panic button

A panic button can be fixed on a wall and his primary function is to trigger an alarm or execute a series of instructions, based on the system configuration.

A panic button may also be carried around in the form of a portable remote controller.

BoPET cable

BoPET (Biaxially-oriented polyethylene terephthalate) is a polyester film made from stretched polyethylene terephthalate (PET) and is used for its high tensile strength, chemical and dimensional stability, transparency, reflectivity, gas and aroma barrier properties, and electrical insulation.
A variety of companies manufacture BoPET and other polyester films under different brand names such as Mylar, Melinex, and Hostaphan. Because they are more straightforward names, brand names are usually referred to and used over the official BoPET name.

Biometric Terminals for Access Control.

Numerous types of Biometric Terminals for Access Control are available on the market.

- Fingerprint reader or Finger vein reader (fingerprint/finger vein reader)
- Iris reader
- Facial recognition reader
- Palm reader or Hand Geometry reader, Etc.

A guide on the pros and cons of each technology is available later in the book under the "to go further with biometrics section."

Card system – (Card reader, contact, and contactless cards)

Although the purpose of this book is primarily biometric technologies, the following paragraphs will cover the topic of card technology to understand how cards are used today and especially how they are used in conjunction with biometric technologies when regulation or extra security features require it.

Many systems originated at a time when biometric readers were not readily available.

Understanding the past can often be useful to understand why things are done a certain way today.

Cards technologies are available as "contact" or "contact-less" and are deemed to be either "standard" or "Smart."

"Smart" in Smartcard refers to the card's storage memory and processing power that is lacking in standard cards. It is often using an extra chip or CPU. Around 1975, Roland Moreno, a French engineer, designed the first Contact Smartcard. Cards can sometimes be referred to as "tag," "credentials" or "token" and so for ease of reading, the terms are used equally and interchangeably in the next few sections.

Contact card system

A contact smart card is usually available in a form similar to the familiar credit card format and contains an embedded smart chip. The user inserts the card into a contact card reader slot. The card reader will then read the UID[3], or CSN[4] contained in the memory on the chip of the card and proceeded according to the setup of the card reader or the access control system. For instance, the system can grant access, deny access or request more information such as a secured code embedded in the memory of the chip (PIN Code, etc.).

The memory of the card depends on the type of cards used; the more the memory, the costlier the card, so it is essential to select a card with enough memory for current and future use.

Swipe or Magnetic Card

Swipe or Magnetic Cards (also called "Mag Stripe") are similar to the Contact Smart Card except that there is no embedded chip but a magnetic stripe at the back of the card.

Nowadays, swipe cards used in access control is declining due to its poor security feature and small memory. It is extremely easy to copy the magnetic stripe of the card, and the tiny storage capacity of the strip limits the number of advanced feature one can use on the system.

TWIC (Transportation Worker Identification Credential)

---

[3] UID is the User ID number of the card
[4] CSN is the Card Serial Number

The TWIC program provides a tamper-resistant biometric credential to maritime workers requiring unescorted access to secure areas of port facilities, outer continental shelf facilities, and vessels regulated under the Maritime Transportation Security Act of 2002, or MTSA, and all U.S. Coast Guard credentialed merchant.

Dallas Tags

A Dallas tag is a plastic tag with a battery using 125Khz frequency.

Other card systems

Over the years, many different types of contact cards or tags were released. There are many other contact card technologies, but this is not the purpose of this book.

Contactless (Proximity) & Contactless Smartcard system

A contactless card or RF (Radio Frequency) system contains two components:

- A contactless card reader ("Transceiver")
- A contactless card ("Emitter")

A contactless card is composed of three components:

- The plastic body of the card,
- A memory chip,
- An antenna rolled in a coil around the card body

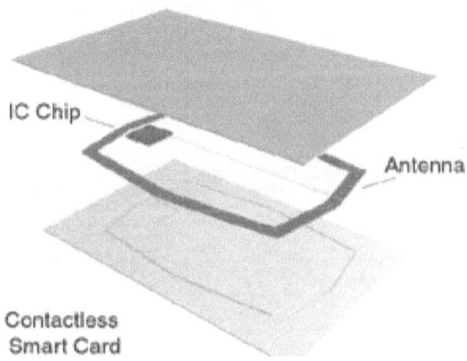

A contactless card can either be "Active" (with embedded battery) or "Passive" (with no embedded battery). The main difference between an "Active" and a "Passive" tag is the attainable reading and emitting range. An Active Card (or tag) automatically broadcast its signal whereas a passive tag will not start emitting until it is powered through a process called "resonant energy transfer."

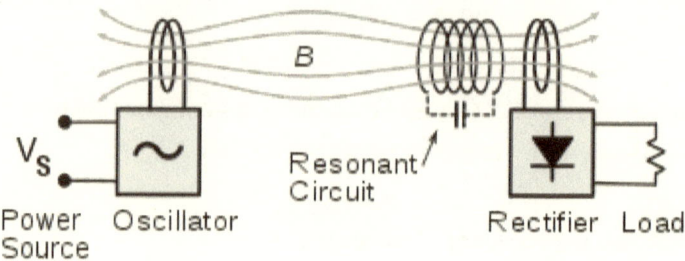

"Resonant energy transfer" is also the principle used for wireless charging. When a card ("the emitter") enters through this energy field, it generates power through induction, and the card accumulates energy through its antenna. This energy is then used to activate the embedded chip. Through this operation, the signal is modified, and the information is transmitted back. In case of a Proximity card, an identifier (User ID) is transmitted back.
In case of a smart card, more information can be transmitted, depending on the type of card and the embedded memory associated with this card. It is, for instance, possible to read an encoded fingerprint template from the smart card and transmit to a biometric reader for matching.

The Transceiver (the reader) listens to and interprets the communication. The table below lists some cards.

Proximity - 125Khz	
Name	**Comments**
Prox®	
Smartcard – 13.56Mhz	
Name	**Comments**
iCLASS®	Seos®, SE,
LEGIC®	
MIFARE®	Classic, Plus, Ultralight, SAM AV2
MIFARE® / DESFire®	EVO1 & EVO2

Contactless smart cards are specified under the ISO/IEC 14443 and the ISO/IEC 15693 OR ISO/IEC 18000 standards. This topic goes above the scope of this book.

Other Contact & Contactless Smartcard

Cards are part of many applications used daily worldwide.
Banking cards are an excellent example.
The financial industry standards such as EMV Certification regulates the specifications of those cards.
EMV stands for "Europay, MasterCard, VISA" and is a standard that manufacturer must adhere to if they want to interact with any financial system or product interacting with those cards (Point of Sale terminals, ATM, etc.). Those standards evolve from time to time with industry requirement such as the use of biometrics, encryption requirements, etc.

For more information: https://www.emvco.com

Biometric Readers mixed modes

Biometric readers features can be used isolated or in conjunction with each other:
Biometric Reader database (1: many)
User ID keyed in on a keypad + Biometric feature Verification from the DB record (1:1)

- *Example:* User ID 1245 – Verification with template 1245 in the database.

Card identifier (UserID) + Biometric feature Verification from the DB record (1:1)

- *Example:* Card identifier = 1245 – Verification with template 1245 from the reader database

Card content (Template on card) + Biometric feature Verification against the card template (1:1)

All these combinations can be combined with a 3rd-factor request for example, a pin code.
The reasons to use these features in combination are usually for enhanced security or because some biometric features are not usable with some individuals (cf. section on Failure to Enrol).

Extra features

Facial detection

Facial detection is often confused with facial recognition.
Facial detection is the ability for a reader to detect if a person is present or not during a transaction. The reader can also store the face capture for audit purposes. The mode can usually be:

- Disabled: The reader does not try to detect a face or to take a picture. No log kept.

- Enabled (best effort): The reader will try to detect a face but proceed with the transaction even if no face is detected. The system saves a picture in the log.

- Enabled (forced mode): The reader will look for a face and will decline access unless a face is detected (and the person is authorised). A picture will be kept in the log file whatever the result.

GPI/GPO

GPI and GPOs respectively stand for General-Purpose Input and General-Purpose Output.

GPI/GPO are an uncommitted digital signal pin, configurable during the system integration/installation.

They are used in many different applications only limited by timing and power.

GPI/GPO typically employ standard logic level ("let the current pass through or not") and cannot supply significant power to run additional devices. However, they can be linked to a third-party device that itself can supply current to a power over devices.

RS Communication

"RS" stands for Recommended Standard and is a standard designed for serial communication.

Serial Communication is a standard of communication which transmits one bit at a time (1 or 0) by changing its electrical state.

Serial communication comes in different versions such as RS-232, RS-422, RS-485 which are the most widely used serial communication protocol in the Access Control industry.

This book does not cover other serial communication protocols such as RS-423 (a high-speed version of RS422 but with unbalanced signalling), RS-449 (a hybrid between RS-232 and RS-422 - Deprecated), and similar.
RS-232, RS-422, and RS-485 communication can be encrypted.

RS232

RS-232 was introduced in the 1960s and is usually recommended to use with a cable no longer than 15 meters and the serial communication often connect through a DE-9M (also called DB9) type connector as per the picture below. It has nine pin connections, but depending on the application, a minimum of 2 pin connections are required (data and ground). Alternatively, a DB25 connector can be used (with 25 pins), but only 9 of the pins are in use (8, 3, 2, 20, 7, 6, 4, 5 and 22). Here is a brief description of each of them:

- Pin 1: CD or Carrier detect
- Pin 2: RxD or Reception
- Pin 3: TxD or Transmission
- Pin 4: DTR or Data Terminal Ready
- Pin 5: GND or Common Ground
- Pin 6: DSR or Data Set Ready
- Pin 7: RTS or **Request To Send**
- Pin 8: CTS or **Clear To Send**
- Pin 9: RI or Ring Indicator

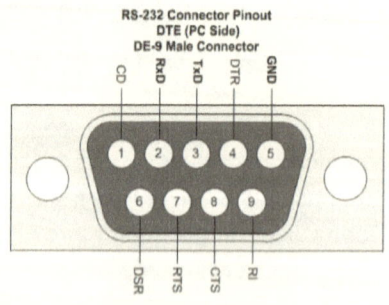 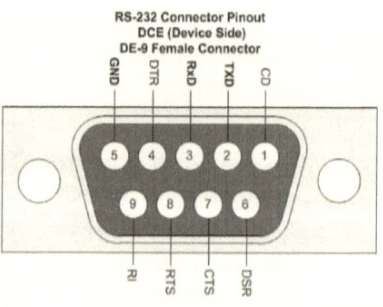

The electrical status changes as follows per the table below.

Data circuits	Voltage
0 (space)	+3 to +15 V
1 (mark)	-15 to -3 V

This standard is intended for use with speed up to 20,000 bits[5] per second (90Kbit/s at 15 meters) and for a point-to-point connection (one device at a time) in half-duplex or full-duplex communication.

RS-422

RS-422 is some "Super RS-232" based on what is called differential signalling. Instead of using only one transmission and one reception pin (Tx/Rx), RS-422 differentiate between the positive and negative signals (Rx+/Rx- and Tx+/Tx-).
The main benefits over the RS-232 are the length which is not limited to 15 meters and the higher speed of transmission.
The speed of transmission degrades with the length of the cable reach a maximum of 10 Mbit/s at 12 meters (40ft) to 90kbit/s at 1200 meters (4000 ft) in half or full duplex as with RS-232).
The connection is made through 5 pins as below and uses the following:

- GND: Common Ground
- Y / Tx+: Transmission (positive)
- Z / Tx-: Transmission (negative)
- B / Rx-: Reception (negative)
- A / Rx+: Reception (positive)

---

[5] Small computer unit – see definition in the definitions section

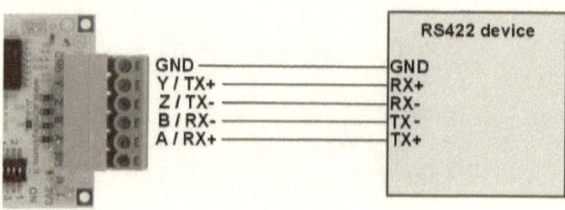

An RS-422 converter can be used on an RS-232 connection to extend its possibility.

RS-422 shares the same point-to-point connection limitation as RS-232 but allows up to 10 devices to receive the signal transmitted by one device whereas RS-232 can only send to one device.

The RS-422 standard does not define a specific type of connector. Usually, it can be a terminal block or a DB9 connector.

The pinout of the RS-422 connector depends on the manufacturer of the device, so it is advisable to refer to the documentation of each manufacturer before installation and configuration of the products.

RS-485

RS-485 is a standard derived from RS-422 but offer a few advantages over RS-422.

It allows for multi-point communication through two wires (for half-duplex) and four wires (for full-duplex), and the maximum speed and length are the same as with the RS-422 specifications.

The maximum number of simultaneously connected devices is 32 and can go up to 256 when making use of repeaters (devices that repeat and amplify the signal).

The RS-485 standard does not define a specific type of connector. Usually, it can be a terminal block or a DB9 connector.

The pinout of the RS-485 connector depends on the manufacturer of the device, so it is advisable to refer to the documentation of each manufacturer before installation and configuration of the products.

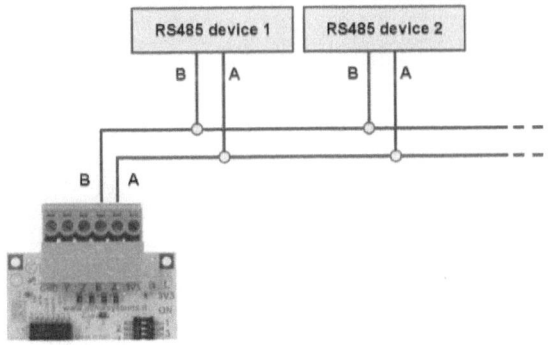

In RS-485, readers connection allows Daisy Chains (one after the other).

Data/Clock

Data/Clock is a mode of communication relying on two wires or lines of communications.
The first line is called "Data" and is used to send Data to a device such as a control panel.
The other line is called "Clock" and oversees synchronisation of communication.
In other words, the "Clock" line acts as a metronome and beat the measure. The receiving device will read transmitted data every time the "Clock" line requests it.

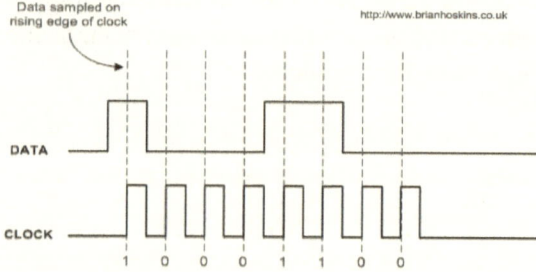

Tamper switch

Within Access Control context, a tamper switch is usually similar to a push button embedded in a device. During the device installation, the switch pushes against the surface on which the reader is placed (a wall, a metal plate, etc.) and get into an engaged status.
If the reader is removed from the surface or opened, the switch disengages and triggers a signal which the device can interpret as per its configuration (alarm, siren noise, send a signal to a third device, etc.)

In most devices, the tamper switch must be activated in the parameters to be able to send the signal, or it will not function.

It is advised to keep the tamper switch deactivated during installation and if it is malfunctioning.

USB connection

USB stands for "Universal Serial Bus" and is an industry standard developed in mid-nineties (1994) specifically for connections between computers and their peripherals.

Before USB connections, computers relied on serial and parallel communication.

At the time of publishing, USB standard has version 3.2 (speed up to 20Gbit/s), and a Wireless-USB version is under work but not yet widely available.

USB also propose different connector types (A, B, C) and allows charging capabilities from 5V,1.5A to 20V, 5A in the latest version.

For Access Control purpose, USB flash drives are usually used to configure the readers via predefined scripts which can be interpreted by the reader. It can also be used to collect information from the reader or load files on the reader (audio clips, picture, video clips, etc.).

Secured connection

A secured connection means an encrypted connection by one or several security protocols (method) to ensure the security and privacy of data exchange between two or more locations.
At the time of publishing this book, many protocols exist, but the most commonly used in access control security network are probably TLS (Transport Layer Security) and SSL (Secure Socket Layer).
Without a secured connection, data circulate in a readable and usable format on a network and can be hacked and exposed to malicious attacks.
If a network is going to be accessible from the outside (WAN, etc.) the use of a Firewall[6] is highly recommended.

Wireless Connectivity

Wireless technology offers a growing range of connectivity option. Readers are embedding them standard or as a plug-in option through external modules. (USB WIFI Dongle for instance).
They all have their purposes and irremediably, their pros and cons. Wireless should however not be considered as a complete replacement of wired solution but as a complement.

---

[6] A firewall is a security device which is screening inbound and outbound connections based on business and security rules. This is handled by IT department.

Bluetooth

Bluetooth is a low-power consumption wireless technology to transfer data between two devices nearby, and there is no need for the two devices to be in the line of sight. Bluetooth speed is up to 24Mbit/s.
Bluetooth uses what is called frequency hopping technology and is, therefore, less inclined to interferences from other wireless technologies. Although it was initially designed for close-proximity, Bluetooth typical range is 10m and with a maximum range up to 100m.
In Access Control, the ranged used is a few metres at most. A Bluetooth reader can be embedded to read information on a device (cellular phone for instance) and replace the need of a separate card. Everybody has a cell phone today so to use it as a card replacement can make sense. The company can save the cost of card issuance and replacement, and anybody using a personal cell phone will take greater care of it than to a simple card.
Some questions remain however when designing a system:

- What if the user forgets his cell phone somewhere?
- What if the cell phone battery is flat?
- What if the user does not want to use his cell phone?
- What about security issues? (signal copying, scramble, etc.)
- What if the phone is being used for a call at the same time (is it convenient to interrupt the call each time access is needed?

Although security is good, it can still be hacked by someone with the right skills do to so
Another use can be a connection between a reader and a third-party device (tablet, laptop, etc.) for configuration and log download for instance.

NFC

NFC stands for "Near Field Communication". Data transmission via NFC is based on Radio Frequency Identification (RFID) standards and uses the same 13.56MHz frequency.
NFC is a low-power communication protocol (even lower than Bluetooth). Both technologies are very similar and therefore interrogations questions related to Bluetooth would probably be the same with NFC.
NFC, however, has a limited range of 4cm and has a maximum speed of up to 424 Kbit/s.
It is not unusual therefore to use NFC and Bluetooth together.
NFC starts the communication and tells both devices what to send and receive. Bluetooth then takes other to send the relevant data.

Wi-Fi / Wi-Fi-Direct

Wi-Fi is a wireless technology using radio frequency to connect devices through an access point (Wi-Fi router). It is the most popular form of Wireless Communication due to its speed, its range and its ease of use.
Wi-Fi specifications are defined by the IEEE [7] Standard "IEEE 802.11".
This standard has several version ranging from 802.11a to 802.11az (planned for 2021).
Each version contains much improvement and can save much money in term of wiring.
802.11a came out in 1999, allows up 54Mbit/s transmission speed for a range of between 35m or 115ft indoor to 120m or 390ft outdoor.
802.11n was released ten years later in 2009, allows up 600Mbit/s transmission speed for a range of between 70m or 230ft indoor to 250m or 820ft outdoor.

---

[7] Institute of Electrical and Electronics Engineers

Those figures are very theoretical though and are largely affected by external factors such as walls, water, interferences (electrical but also from other wireless devices).
Wi-Fi communication must be secured to avoid data interception and hacking.
Wi-Fi is a convenient technology but is not always the solution and if stability and security of communications is the number one priority Wi-Fi should be considered carefully.
When a device connects to a Wi-Fi network, it gets a local IP address and is an integral part of the IP network which includes Wired and Wireless networks.
To balance the original limitation of the Wi-Fi design which only allows connections through a Wi-Fi router, the IEEE released Wi-Fi Direct. Wi-Fi Direct allows two devices to establish a direct Wi-Fi connection without requiring a Wi-Fi Router.
However, to take advantage of all those features both devices must be compatible with the same Wi-Fi versions. In other words, if a device is compatible with 802.11 a, g & n but the second device is only compatible with 802.11a, the communication will only be as fast and as far as 802.11a allows even though 802.11n should allow the first device to transmit faster and further.

3G/4G/LTE

3G, 4G and LTE are all mobile communication technologies. While they are wireless technologies, they are all long-range technology used by the mobile operators to exchange data with their customers' mobile terminals (cellular phones, tablets, etc.).
The numbers "3" and "4" respectively stand for the third and fourth generations of the technology.
LTE stands for "Long Term Evolution", and the full name is 4G LTE and is in theory up to 10 times faster than 3G.
The advantage for mobile communication over standard wireless technologies such as Wi-Fi or Bluetooth is that there is no limitation of a site

communication range so long as the user is within the coverage of a mobile operator. A reader can therefore be on-site and managed remotely from an office located several hundred kilometres away.

The main disadvantage is that the need for a bit more hardware and settings to get going:

- Device accepting a SIM Card (via an onboard router, a dongle[8] or connector).

- A SIM Card and a contract (or at least prepaid service) from a mobile operator.

- Set up a private APN (Access Point name) or connect to a public APN.

Depending on the device trying to connect, and the user requirements, some customisation coding might be required.

Unfortunately, mobile communication data is not free, so without an unlimited usage contract, the bill might become quite costly if a large amount of data is transmitted every month.

The other limitation is that if there is no network coverage, the solution will not function properly.

Turnstile

A turnstile is a rotating gate which is activated by an internal relay. Two type of turnstiles exists namely basic turnstile and advanced turnstile.

Traditional turnstiles are fitted with a standard relay and are therefore only able to open or close and let the turnstile move into one direction only (going in or out).

More advanced turnstiles can open in one direction or the other, based on logic and external device. For example, if a user is entering a building with

---

[8] A dongle is a device (USB, Serial or Parallel) used with or without a license to protect a software.

his fingerprint, the turnstile should be able to detect and open only in the entering position and vice-versa.

Some turnstile embeds even more advanced logic and allows for a precise status monitoring.

In other words, if a user clocks in but doesn't go all the way through the turnstile, the device would be capable of indicating in its event logs that the user was granted access but did not go through the turnstile. This feature is handy on site where users could be tempted to clock in but never go through the turnstile and come back later to clock-out.

Speed Gates

Speed gates are used to created delimited areas, designed to direct people in the desired direction, as fast and conveniently as possible. Speed gates can be closed with plexiglass barrier and open via a relay triggered upon authorisation via biometrics, card or remote control.
Some speed gates operate without any mechanical barrier and rely solely on sensors to detect if the perimeter is breached. The use of this technology is typically only practical if manned 24hours by security officers in close proximity.

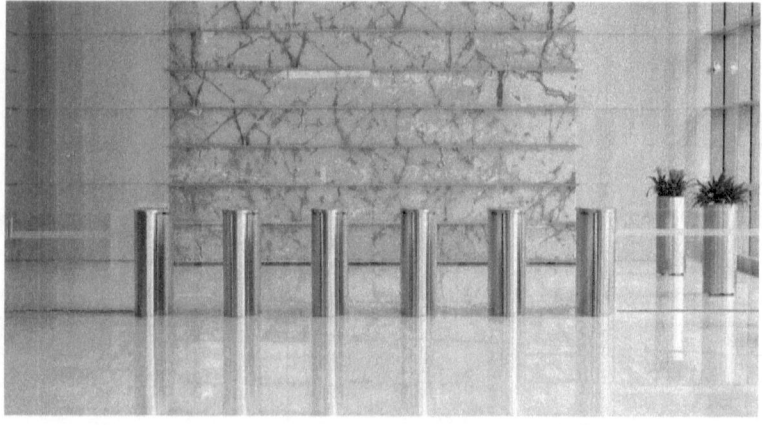

A new trend is currently emerging: "reversed access control".
In reversed access control, the barriers are open until an unauthorised person tries to gain access.
The barriers close as fast as necessary, proportionally to the speed at which the unauthorised person approaches to prevent access.
To avoid tailgating, infrared beams are usually placed on each side of the gates as well as on top of the gates. Infrared systems can prevent tailgating and at the same time ensure that no one can walk through the gate from the side instead of someone who just authenticated himself.

Secured vs non-Secured Area

There can be no Access Control without differentiating between two types of areas:

1. Unsecured area
2. Secured area

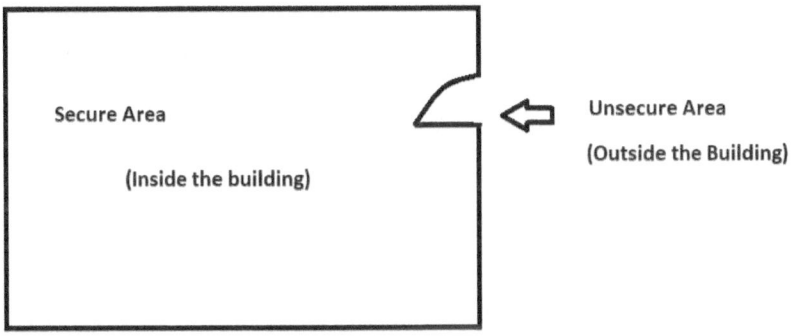

The secured area is resulting in the implementation of the access control system.
The unsecured area is the area leading to the secured area. Access Control is installed between the two areas.
It is essential to keep as little security equipment as possible installed in the unsecured area to minimise the possibility of anyone tampering with the system.
In a typical access control installation, the only gears accessible from the unsecured area should be those that with whom users need to interact.
However, during the design phase of a security system many criteria such as budget, space, level of security required, will influence the system implementation.
It is the role of the consultant, system integrator or installer to advice the end-user of the best trade-off possible but ultimately, the choice remains with the end-user. All types of implementation have pros and cons as described in the following section.

Access Control deployment types

Access control implementations can take many different approaches depending on the level of security required and the budget.

Standalone

A standalone implementation of access control is the most basic implementation possible.
It consists of installing a reader on the wall next to the door to control and to connect this reader directly through its internal relay to a door locking mechanism such as a striker or a magnetic lock (covered later in this book).
It is quick to install and reasonably quick to set up. It is cost effective but in return not very secure and only adapted to use on a small-scale system such as a couple of doors and for a couple of people only, due to the limited administration possibilities.
Each user must enrol on each reader and the event logs collect must be done directly on the reader itself, usually via a USB flash drive or similar.
The level of security is minimal as anybody with access to the cabling could replace the relay and short the connection to open the door.
Access to the cable could happen by cutting the cable trunk running along the wall to the reader.

It could be even worse if the cable is directly glued to the wall without protection (terrible practice).
If no cable is exposed on the wall, it is equally easy to access it by removing the reader from the wall and accessing the cables at the back. In this case, it is recommended to activate the reader's internal tamper switch when available. An alarm will go off upon the removal of the reader from the wall.
It is strongly advised to avoid this type of setup unless if it is used for convenience only and where security is not critical as such.

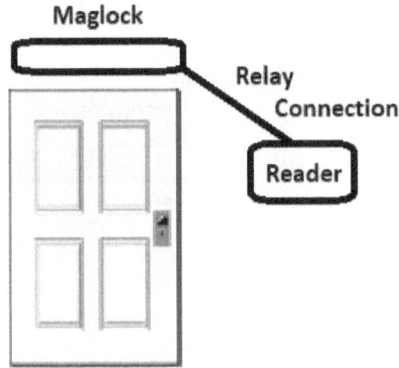

Enhanced Standalone

An enhanced standalone solution extends a standalone solution in the sense that it connects directly onto the door locking mechanism, but it is at the same time connected to an IT network through TCP/IP or similar technology which allows for a more flexible administration of the system.

For instance, enrolment happens on a dedicated enrolment station, and the template can be deployed in the readers directly from the computer. There is, therefore, no need to enrol all the users on each reader.

As the readers are networked connected, it is also much easier and convenient to collect the event logs from each reader.

The level of security, however, is inherited from the standalone typology and is still minimal.

This setting is still not recommended unless for convenience purposes only.

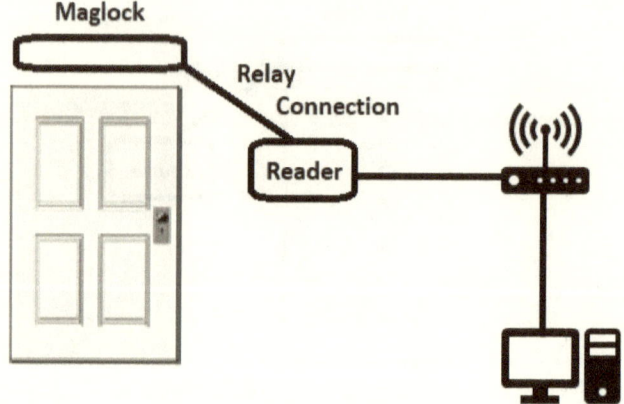

Standalone with Secured Relays

The next level of security, whether a standalone or an enhanced standalone system is set-up, is to make use of a more secured relay called intelligent relay or "e-relay". The reason why those relays are deemed to be intelligent is that they are waiting for a specific code or signal to be sent from the reader to open the door.

One cannot, therefore, open the door just with simple by-passing methods as described in the case of the standalone or enhanced standalone system.

The secured relay should be installed in the secured zone to prevent exposure to unauthorised access.

The secured relay is however limited to opening upon authentication as advanced rules set-up is not available.

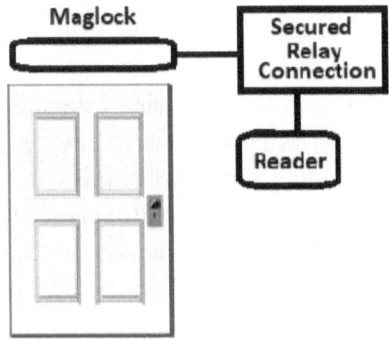

PACS (Physical Access Control System)

In Access Control like in many other industries, standardisation has taken place.
The most known protocol until now has been Wiegand, and although new standards such as OSDP are emerging, it will still take a long time before Wiegand disappear entirely.

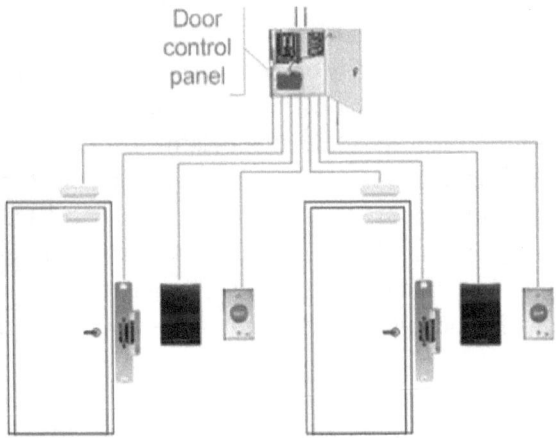

Wiegand

The **Wiegand** interface is a widespread wiring standard in the access control industry. The Wiegand interface originated from the popularity of **Wiegand** effect card readers in the 1980s. It is commonly used to connect a card or biometric reader to an access control panel or controller.
Wiegand transmits information (usually an identifier) through two wires called Data 0 (carry the 0 bit) and Data 1 (carry the 1 bit). The controller records the data as it receives it.
Although Wiegand has been a de facto standard in the industry for so many years (from the early 1980s), it has not evolved to adapt to today' security demands. It is an unsecured standard, and OSDP is more and more recommended to replace Wiegand.

OSDP

Open Supervised Device Protocol (OSDP) is an access control communications standard developed by the Security Industry Association (SIA[9]) to improve interoperability among access control and security products. OSDP v2.1.7 is currently in process to become a standard recognised by the American National Standards Institute (ANSI), and OSDP is in constant refinement to retain its industry-leading position.
OSDP is often seen as the successor of Wiegand protocol and according to the SIA (Security Industry Association of America) offers higher security than other communication protocol through amongst other the support of high-end AES-128 encryption (required in US federal government applications).
OSDP also continually monitors wiring to protect against attack threats and proposes advanced functionalities such as the supports for advance smartcard technology applications, including PKI/FICAM and biometrics. Furthermore,

---

9 SIA – Security Industry Association of America – https://www.securityindustry.org

encryption and authentication are predefined so that no guesswork takes place.

OSDP supports bi-directional communications among devices, and advanced user interface, including welcome messages and text, prompts.

OSDP uses two wires instead of twelve or more which allows for multi-drop installation, supervised connections to indicate reader malfunctions, and scalability to connect more field devices.

OSDP is introducing improved ease of use through audio-visual user feedback mechanisms provide a rich, user-centric access control environment.

The OSDP specification is currently recommended when TCP/IP, USB, or other conventional protocols do not lend themselves to the application.

The OSDP specification is extensible to IP environments, and the OSDP WG is working on deploying OSDP over IP soon.

For more information, please visit the SIA website.

ONVIF Profile A[10]

A fixed set of functionalities describes an ONVIF profile through several services that are provided by the ONVIF standard. Several services and functionalities are mandatory for each type of ONVIF profile. An ONVIF device and client may support any combination of profiles and other optional services and functionalities.

Profile-A is the ONVIF Profile dedicated to Access Control. It is not currently very spread but might be developing further in the future.

For further information, visit the Onvif Website.

Full integration

PSIM (Physical Security Information Management)

---

[10] - Onvif Website: https://www.onvif.org/profiles/profile-a/

The PSIM is an all integrated platform which purpose is to ensure that all components of a security system work in sync to enforce security rules. PSIM involves information management. Information in this context can be user identity, access control right, sensitive areas to monitor, trigger action of a component of the system if an event is detected. To that effect, PSIM platform works through electronic controllers installed at strategic places. There are two components of controllers: Main Controllers (MC) and Door Controllers (DC).

MCs and DCs work similarly to a client/server architecture in IT where a client will log onto a server, follow a certain number of rules and come back to the server for further information when necessary.

Main Controllers store all Access Control rules and deploy those rules through to the appropriate Door Controllers.

Door Controllers will wait to receive electronic input or feedback from devices, analyse the feedback and trigger the next action based on the defined rules.

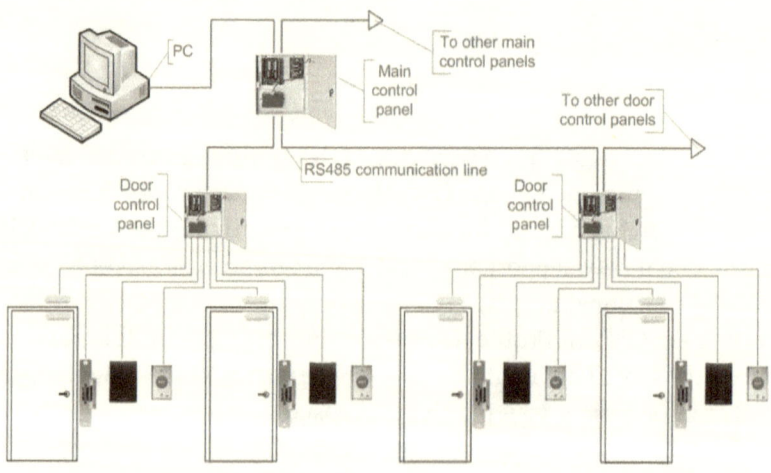

For instance, if someone tries to gain access to an area where he does not have the right to be, the system may instruct a CCTV camera to focus on the zone where it is happening and send a signal to an operator to instruct him to have a look and decide if an intervention is needed.

PSIM platforms usually embed the possibility to enrol biometrics for its users directly from the platform.
PSIM manufacturers work together with component manufacturers and use SDKs[11] to integrate those components seamlessly into their platform.
The advantage is that operators do not need to use different software to manage the various components.
The drawback, however, is that it can sometimes take a bit of time between the release of a new component to the market and the full integration into a PSIM System.

Multi-authentication

A controller may be programmed to request further security tokens such as a card or a second card, a pin-code or anything else as programmed in the system.
Although it is also possible to program those onto some readers directly, it might be useful to delegate this to the controller.
For instance, to ask for a second means of verification randomly or different from one user to the other. The Controller would ask a user for his fingerprint and the presentation of a card and the next time, it might ask for the fingerprint and a pin-code or even for the three in a specific order.

Multi-authentication is considered to be more secure than a single authentication system as on top of the traditional access control requirement of "who you are?", "where you are?" and "when you are?", it relies on two or more of the following:

---

[11] - Software Development Kit

- Biometrics - Who you are?
- A physical token - What you have?
- Code or password – What you know?
-

Anti-passback

Anti-passback is a simple but powerful feature of a PSIM system. It ensures that somebody cannot enter the facility twice if there is no record of them exiting after the first entry. Conversely the system can be configured to not allow an exit if an entry for the person has not been recorded.
The goal is to deter employees from badging someone else through, on their own credentials[12] (e.g. a visitor or another employee). In other words, any given employee cannot have two successive entries, or two successive exits at any access point.
The efficiency of anti-passback is however largely dependent on the type of access control implementation and requires a system that ensures only one person at a time can go through. That is, for instance, a full-height turnstile.

Dual or joint Authorization

In some cases, a business might not want anybody to enter a restricted area alone also called a "NO-ALONE Zone" or "2 Man Rule".
For instance, a bank might not want anybody to be able to access the vault alone. In this case, if somebody tries to enter the vault, the information would be sent to the Door Controller which would in return ask for a second person to authenticate before deciding to open the vault or not.

Zone counting

---

12 - Credentials: Access Control

During an emergency evacuation, it could be necessary to determine in real-time who is in a zone or not.

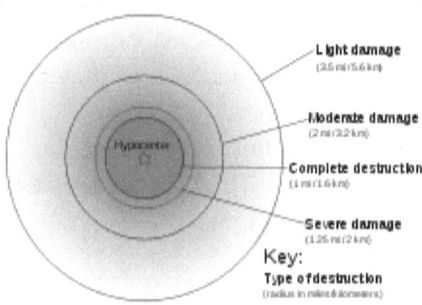

In the mining industry, blasting operations can happen at reduced intervals thanks to PSIM technology. Before PSIM, the lapse between two blasting could be very long due to safety considerations. To ensure no one was left behind, the blasting zone was divided into several areas which had to be evacuated at least x hours before the blast. The count happened manually, and it was a long and strenuous process. With the introduction of PSIM, the lapse between blasting shrunk dramatically. The system counts everyone entering a zone and everyone exiting this zone. It becomes, therefore, easier to see who is where and to see if blasting can go ahead or not. This not only increases efficiency for the mine but also greatly increases safety for the employees.

Device Monitoring

With PSIM, it is possible to monitor component of the system and trigger an event if the component reacts within or out of a specific set of parameters. For instance, if a door remains open for an abnormal period, the system can trigger an alarm or alert an operator and as seen previously, can redirect a camera towards the problematic zone.
Likewise, it is possible to switch the air conditioner on if the temperature rises above a certain level in a specific zone. Sometimes, there is a fine line between PSIM and building automation.

Perimeter control

Perimeter control can happen in many ways. Some are classic like having a barbwire or mesh fencing.
Some are a bit more secured like electric fencing which can also be monitored directly via the PSIM system.
In recent years, with the rise of video analytics, cameras have started to be used to
Although it will not stop intruders, it can alert an operator of trespassing, and if used in conjunction with electric fencing and positioned strategically, this can be extremely efficient as it ensures interception of intruders before they even get to the physical barrier itself. The drawback of video-based solutions is that external factors like sun, shadows, wind, and so forth, sometimes affect performances and make it difficult to set up.
Finally, thermal cameras are available on the market, and although they are more expensive than standard cameras, they can prove to be very efficient to pick up human being during night time.

BSIM (Business Security Information Management) / OHS (Occupational Health and Safety)

BSIM is based on PSIM but goes one step further.
PSIM only involves security rules where BSIM also involves business rules in the process. Through BSIM, a business can make sure that it complies with local rules and regulations.

Compliance is of utmost importance especially for regulations such as those of Occupational Health and Safety (OHS).

Let's take mining as an example. Over the years, regulations have become very stringent to ensure that miners are not taking unnecessary risks and ultimately that lives are saved or bettered.

Induction[13] programs are designed and implemented, and specific permits are needed to enter specific areas.

Furthermore, regular health checks have become mandatory. Not only are these regulations for the well-being of employees, but it has also at the same time become a burden on the employer who must ensure that their enforcement and that proper records keeping.

BSIM can ease the burden of employers by ensuring that only those employees who are up-to-date with the rules can gain access to specific areas and then ultimately incentivise them to abide by those rules as the contrary could have a direct impact on their payslip as we will see under the Time and Attendance section.

At the point of entry, the employee will be requested to identify himself/herself, and that is where the use of biometrics is paramount to ensure identification of the right person.

---

[13] Introduction and training to a specific environment: mining in this instance.

Once the front-end reader identified the employee, it sends the user identifier to the door controller which will check if the user can enter.
In parallel, it will go through a checklist of parameters and will open the door or trigger more actions if necessary.
It could for instance display on a screen a reminder to an employee mentioning that his next medical check-up is due in 30 days and allow him entry. The following days, it would remind the employee that his medical check-up is due in 29 days and so on and so on.
After 30 days, the message would change to "Medical Clearance expired, please report to infirmary", and the employee would not be allowed in until he goes for a medical check-up and his credentials updated. Then only will the employee be able to request access. The same could apply to a driving license or a special permit needed to drive specific vehicles for instance.
The BSIM system is also able to question further components of the system for further clearance upon identity confirmation. For instance, a random breathalyser test can be requested, to only allow sober employees to come in. All transactions are logged for further disciplinary procedures when necessary.

Middleware integration

Middleware is a software which acts as a connection or as a bridge between two different systems.

Sometimes, Access Control Manufacturers do not want or don't have the capacity to integrate biometric systems directly into their platform. In this case, they can make use of a middleware integration proposed by the biometric manufacturer or 3rd party software providers. In this case, the Access Control Manufacturers only need to add a link to the enrolment interface into his solution. Both software synchronises common data in their databases in the background.
The advantage of this solution is that the Access Control Manufacturers do not have to worry about biometric integration every time a new model or a new firmware14 version comes out.

The second advantage is that as opposed to the hard integration, there is no need to capture the information on both systems and the user identifier is automatically the same.
The drawback for the Access Control Manufacturers is that they are dependent on the biometric manufacturers to develop its interface.

ACaaS (Access Control as a Service)

Access Control as a Service allows end-users to pay a subscription fee and get an always up-to-date platform with security patches and the latest features. The end-user saves on upfront investment for the platform but must still bear the hardware and installation cost.

The biggest challenge of ACaaS remains its dependability on the end-user IT infrastructure (and internet connectivity in particular) which must be good enough to support the service.

However, for companies with multiple interconnect site, the problem would be the same with a traditional system, and therefore ACaaS might be an option too.

14Embedded software that the device needs to function

Access control solutions Summary (1/2)

Type of access control	Pros	Cons	Comment
Standalone	• Cheap • Quick to install and setup	• Not secure • Limited access control features	
Enhanced Standalone	• Cheap • Quick to install and setup • A bit more flexible to administrate than standard Standalone	• Still not secure • Limited access control features	
Standalone + Secured relay	• Cheap • Quick to install and setup • A bit more flexible to administrate than standard Standalone • More secure than plain standalone	• Limited access control features	

Access control solutions Summary (2/2)

Type of access control	Pros	Cons	Comment
Hard integration	• Much more secure than standalone • More access control options	• More expensive than standalone solutions • Still basic integration. • Information to be captured in both systems via two different interfaces	For the solution to function, user identifier must be the same on both systems.
Middleware integration	• Ensure that Biometric integration is always up to date	• Can be quite limited in access control functionalities	
Full integration: PSIM / BSIM / OHS	• Very advanced settings • Full business control	• Can be quite costly compared to other systems	

Network

For a long time, IT and Security departments remained separated, and IT departments were denying access to any security system onto their network infrastructure. Long gone are those days and Security is becoming increasingly part of IT's duties.

Therefore, while IT Networks is above the scope of this book, a minimum of information is necessary to understand how IT equipment and infrastructure impact and interact with security systems, especially as IoT[15] is rising.

At least two components can identify devices which connect to an IT network:

- Hardware: Media Access Control or MAC Address.

- This address is unique per device and is set up by the manufacturer of the product and cannot be modified.

- Software: IP Configuration (for Internet Protocol Address).

- Every system has a default IP configuration which can be changed by the system administrator.
-

---

[15] Internet of Things

An IP configuration is the virtual or logical version of a physical address:

	Physical	Logical or Virtual
Address	House number i.e. N#4	IP address i.e. 192.168.1.10
Location	City	Subnet Mask i.e. 255.255.255.0
Gateway	Mail Exchange	Distant Site Gateway address i.e. 66.5.1.202

The Gateway address is the address to which all communication destined to go outside of the local site will use.

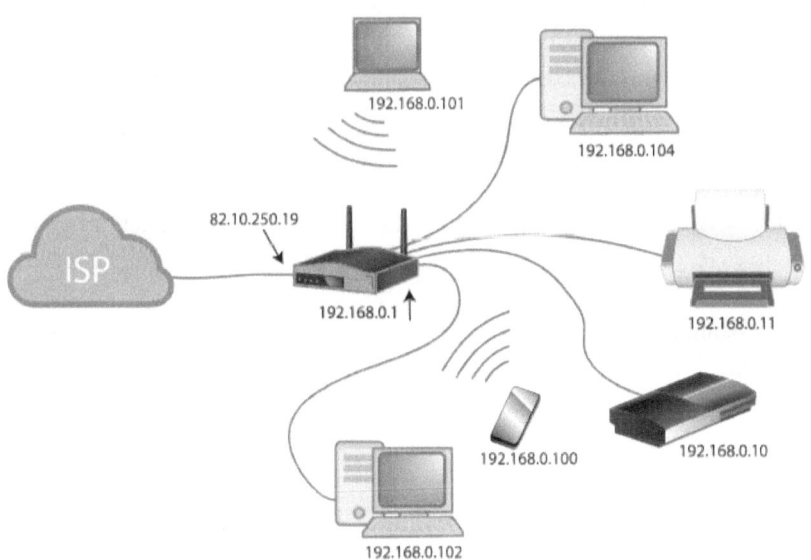

This chapter is deliberately not touching on the advanced concept of "IP addresses Classes", "IPv4 vs IPv6" and so forth as sufficient information is available through the internet if necessary.

POE/POE+

POE stands for Power-over-Ethernet and gives the possibility to power a device connected to a Local Area Network (LAN) so long as the SWITCH it connects to is POE capable.

The main advantage is that power and data are using the same cable and therefore only one cable is required reducing the cost of installation (less cabling, faster to install, etc.).

The original PoE standard provides up to 15.4W of DC power (minimum 44 V DC and 350 mA[3][4]) on each port. However, only 12.95 W is assured to be available at the powered device as some power dissipates in the cable. The updated version of POE, also known as PoE+, initially provided up to 25.5 W of power and the 2009 standard prohibits a powered device from using all four pairs for power.

Standards body IEEE[16] regularly releases new versions specifications of POE with extended power(55W, 90W and more) and bandwidth(2.5Gb, 5Gb, etc).
Some non-standard implementations also exist and may not always be well supported by hardware with standard implementation. Avoid them to avoid issues that may take time and money to be resolved, if ever possible.

---

[16] Institute of Electrical and Electronics Engineers

Access Control installation recommendation

As already mentioned earlier in this book, proper planning is essential to any project.
Here is a list of recommendations for any installation.

Change default passwords.

The default password is probably one of the most important details that should be checked during commissioning of the installation, yet too often overlooked. Way too many installers leave the site before having changed the default password on devices. They are usually not informing the end-user or owner of the system of such password neither.
Not changing the default password is a massive security risk, and it defeats the purpose of investing in very sophisticated and costly access control system.
The password should be kept in a safe place for future access (at least by the end-user) and by the installer with the authorisation of the end-user (especially if an SLA17 is in place).

Pc Security – Make sure PC software is up to date

- Operating System & its patches
- Antivirus
- Manufacturer software
- Etc.

Hardware Firmware

---

17 Service Level Agreement – A contract where a regular service is agreed upon between the service provider and the end-user.

Check and update each device firmware version according to the manufacturer and third-party solution developer's recommendations.

Cables labelling

Labelling of cable on both ends is paramount to world-class installation. Labelling takes a bit more time at first, but it looks much more professional, ensures easy maintenance and will save time in the long run.

Create a Map/diagram of the site

Drawing a map or a diagram of the site with as many details as possible is a good practice. It ensures easy troubleshooting, maintenance and upgrades at a later stage.
The following list is but a few Information that should be on the map.

- Designation of the product (reader, server, enrolment station, door controller, etc.)
- An IP address (IP, subnet, gateway)
- Alimentation type (POE, POE+, 12V, etc.)

Any comments concerning the site and the installation can also be in the diagram.

Leave the site neat

Upon conclusion of the installation, the site should be as neat as possible. It means that it should be clean and that all garbage generated during installation should be removed. If anything has been soiled, it should be cleaned. The end-user should find the premises as neat (if not better) than when installation started. It looks professional and gives a sense of accomplishment. Customers are more likely to call the installer back for more projects if they have the feeling that he knows what he is doing, looks professional, and foremost that the installer seems to care about the customer.

Commissioning of a site is always good practice

Commissioning is the phase where the installation of a site is wrapped up. During this phase, the following actions are taking place:

- The system is thoroughly tested to ensure that it operates within the parameters set in the project specifications and the end-user witnesses that the installations functions according to specifications.

- The customer's team is trained to use the products and perform basic maintenance and troubleshooting.

- If required, the employees/end-users must be trained on how to use the system (including the do's and don'ts is applicable)

- Communicate passwords and any specific information to the customer.

- Propose and finalise a Service Level Agreements (SLAs) if necessary.

- Finally, the customer's sign the project off and the final invoice can be raised and issued.

In case of doubt, ask the manufacturer or their local representation for assistance.
Nobody knows everything, and there is no shame in asking for assistance. Most of the time, you will get assistance and training might even be available.
If the manufacturer proposes a certifications program, jump in. It does make the end-user more confident in your ability to fulfil the job. Between a certified and a non-certified company, the choice will undoubtedly go towards the certified company.

For further information, visit https://biometric.i-learn-online.com

Note about software and firmware update:
Ensure firmware compatibility with each piece of the solution before updating.
Solutions are rarely all updated in real time and therefore, manufacturers need time to update their software. If you upgrade a reader with a firmware which is not certified as compatible with the third-party solution, the system may be at risk to stop functioning correctly.

Industry Standards

As in any industry, the security industry adopted certain standards, and we are covering the most important to know here.

- Intrinsically safe
- Ingress Protection
- IK rating

IP Ratings (Ingress or International Protection)

IP Code or International Protection Marking, also called Ingress Protection, classifies and rates the degree of protection provided against intrusion (body parts such as hands and fingers), dust, accidental contact, and water by mechanical casings and electrical enclosures. It is published by the (IEC[18]). The equivalent European standard is EN 60529.

The first digit stands for Solid particle protection (ranges between 0 and 6).

Level-sized	Effective against	Description
0	—	No protection against contact and ingress of objects
1	>50 mm	Any large surface of the body, such as the back of a hand, but no protection against deliberate contact with a body part
2	>12.5 mm	Fingers or similar objects
3	>2.5 mm	Tools, thick wires, etc.
4	>1 mm	Most wires, slender screws, large ants etc.
5	Dust protected	Ingress of dust is not entirely prevented, but it must not enter in enough quantity to interfere with the satisfactory operation of the equipment.
6	Dust-tight	No ingress of dust; complete protection against contact (dust tight). A vacuum must be applied. Test duration of up to 8 hours based on air flow.

The second digit stands for Liquid ingress protection and ranges from 0 to 9.

[18] International Electrotechnical Commission

Level	Protection against	Effective against	Details
0	None	—	—
1	Dripping water	Dripping water (vertically falling drops) shall have no harmful effect on the specimen when mounted in an upright position onto a turntable and rotated at 1 RPM.	Test duration: 10 minutes Water equivalent to 1 mm rainfall per minute
2	Dripping water when tilted at 15°	Vertically dripping water shall have no harmful effect when the enclosure is tilted at an angle of 15° from its normal position. A total of four positions are tested within two axes.	Test duration: 2.5 minutes for every direction of tilt (10 minutes total) Water equivalent to 3 mm rainfall per minute
3	Spraying water	Water falling as a spray at any angle up to 60° from the vertical shall have no harmful effect, utilising either: a) an oscillating fixture or b) A spray nozzle with a counterbalanced shield. Test a) is conducted for 5 minutes, then repeated with the specimen rotated horizontally by 90° for the second 5-minute test. Test b) is conducted (with a shield in place) for 5	For a Spray Nozzle: Test duration: 1 minute per square meter for at least 5 minutes Water volume: 10 litres per minute Pressure: 50–150 kPa For an oscillating tube:Test

		minutes minimum.	duration: 10 minutes Water Volume: 0.07 l/min per hole
4	Splashing of water	Water splashing against the enclosure from any direction shall have no harmful effect, utilising either: a) an oscillating fixture or b) A spray nozzle with no shield. Test a) is conducted for 10 minutes. Test b) is conducted (without shield) for 5 minutes minimum.	Oscillating tube: Test duration: 10 minutes, or spray nozzle (same as IPX3 spray nozzle with the shield removed)
5	Water jets	Water projected by a nozzle (6.3 mm) against enclosure from any direction shall have no harmful effects.	Test duration: 1 minute per square meter for at least 3 minutes Water volume: 12.5 litres per minute Pressure: 30 kPa at a distance of 3 m
6	Powerful water jets	Water projected in powerful jets (12.5 mm nozzle) against the enclosure from any direction shall have no harmful effects.	Test duration: 1 minute per square meter for at least 3 minutes

			Water volume: 100 litres per minute Pressure: 100 kPa at a distance of 3 m
6K	Powerful water jets with increased pressure	Water projected in powerful jets (6.3 mm nozzle) against the enclosure from any direction, under elevated pressure, shall have no harmful effects. Found in DIN 40050, and not IEC 60529.	Test duration: at least 3 minutes Water volume: 75 litres per minute Pressure: 1000 kPa at a distance of 3 m
7	Immersion, up to 1 m depth	Ingress of water in harmful quantity shall not be possible when the enclosure is immersed in water under defined conditions of pressure and time (up to 1 m of submersion).	Test duration: 30 minutes - ref IEC 60529, table 8. Tested with the lowest point of the enclosure 1000 mm below the surface of the water, or the highest point 150 mm below the surface, whichever is deeper.

Level	Protection against	Effective against	Details
8	Immersion, 1 m or more depth	The equipment is suitable for continuous immersion in water under conditions which shall be specified by the manufacturer. However, with certain types of equipment, it can mean that water can enter but only in such a manner that it produces no harmful effects. The test depth and duration is expected to be higher than the requirements for IPx7, and other environmental effects may be added, such as temperature cycling before immersion.	Test duration: Agreement with Manufacturer Depth specified by the manufacturer, generally up to 3 m
9K	Powerful high-temperature water jets	Protected against close-range high pressure, high-temperature spray downs. Smaller specimens rotate slowly on a turntable, from 4 specific angles. Larger specimens are mounted upright, no turntable required, and are tested freehand for at least 3 minutes at distance of 0.15–0.2 m. There are specific requirements for the nozzle used for the testing. This test is identified as IPx9 in IEC 60529.	Test duration: 30 seconds in each of 4 angles (2 minutes total) Water volume: 14–16 litres per minute Pressure: 8–10 MPa (80–100 bar) at a distance of 0.10–0.15 m Water temperature: 80 °C

Popular ratings in Access Control are IP65, IP66, and IP67.

Intrinsically safe

Intrinsic safety (IS) is a protection technique for safe operation of electrical equipment in a hazardous environment. It, in principle, limits the electrical and thermal energy available for ignition. Areas with dangerous concentrations of flammable gases or dust are found in applications such as petrochemical refineries and mines. High-power circuits such as electric motors or lighting cannot use intrinsic safety methods for protection. Consideration for IS must start at conception phase, and that is a synonym of cost increase. That is the reason why very few manufacturers include IS in their products unless they are specifically designed to target an IS sensitive market.

IK Rating

IK ratings are defined as IKXX, where "XX" is a number from 00 to 10 indicating the degrees of protection provided by enclosures against external mechanical impacts. The **IK rating** scale identifies the ability of an enclosure to resist impact energy levels measured in joules (J)

Rating	Protection level
IK00	Not protected
IK01	Protected against 0.14 joules impact. Equivalent to impact of 0.25 kg mass dropped from 56 mm above impacted surface.
IK02	Protected against 0.2 joules impact. Equivalent to impact of 0.25 kg mass dropped from 80 mm above impacted surface.
IK03	Protected against 0.35 joules impact. Equivalent to impact of 0.25 kg mass dropped from 140 mm above impacted surface.
IK04	Protected against 0.5 joules impact. Equivalent to impact of 0.25 kg mass dropped from 200 mm above impacted surface.
IK05	Protected against 0.7 joules impact. Equivalent to impact of 0.25 kg mass dropped from 280 mm above impacted surface.

IK06	Protected against 1 joules impact. Equivalent to impact of 0.25 kg mass dropped from 400 mm above impacted surface.
IK07	Protected against 2 joules impact. Equivalent to impact of 0.5 kg mass dropped from 400 mm above impacted surface.
IK08	Protected against 5 joules impact. Equivalent to impact of 1.7 kg mass dropped from 300 mm above impacted surface.
IK09	Protected against 10 joules impact. Equivalent to impact of 5 kg mass dropped from 200 mm above impacted surface.
IK10	Protected against 20 joules impact. Equivalent to impact of 5 kg mass dropped from 400 mm above impacted surface.

Logs

For every transaction, biometric readers keep an event log.
The format of this log is usually something very similar to:
User_ID, Date, Timestamp, Reader, Direction, Operation, result.
User ID = Unique identifier of the user in the system (can be an employee number).
Date = Precise day when the transaction occurred, e.g. 14/07/2017
Timestamp = Precise time of the transaction, e.g. 18:23
Reader = Name allocated to the reader, e.g. "Front door"
Direction = In which way was the user going, e.g. "IN" or "OUT"
Operation = Usually "Clock in", "Clock out", "Off Duty", "On Duty"

The result can be one of the following:

- Identified, which implies that only the biometricsbiometric feature was presented to the reader.
- Verified, which implies that a token was also presented (card, code, etc.)
- Unidentified, in which case the User_ID should be empty as the user is not identified
- Access Denied, which implies that the user is known but is at least not authorised to enter this area at that time and date).

Other information can be part of the log itself or can be retrieved from the system, e.g. name, surname, etc.

Those logs are useful for two main purposes:

- Security audit & investigation purposes – to find out who was in or near a specific zone at a specific time and date.

- Time tracking with Time and Attendance systems which is discussed in the next chapter.

6. Time and Attendance systems (TNA)

Time and Attendance systems (T&A) have historically been used to keep documented records of who was present, at what place, and from a starting time to an end time. T&A used to be done manually through a registrar before moving to a clocking machine with paper and later plastic cards before logically moving towards biometrics.
T&A systems are usually used to figure out how long did an employee work during a specified period. From this information, the system can deduce if an employee worked overtime, under time or was absent.
TNA systems rely on event logs generated by readers to report on working hours.
Biometric technologies help prevent the following:

Ghost employees

A ghost employee is an employee who exists in the payroll system but isn't a real employee of the company. Ghost employees happen when somebody creates a profile in the system and use the card allocated to this profile to clock him in and out daily. The illegitimate wage later allocated to this ghost employee is paid over to the account of the fraudster.
A ghost employee can also be a former employee who is no longer working for the company whose profile is still present on the system. In this case, the banking details are merely updated.

Buddy Clocking

Buddy Clocking is when somebody clocks for an employee who is not present at work.
That could be as simple as an employee "taking a day off" or as sophisticated as employees alternating their presences so that they can work a day out of two or have several jobs at the same time. In some cases, employees could be

"renting their job" at a lower cost than what they get paid and create a business out of it.

Time Loss

Another issue is time lose which very often is underestimated. Employees getting in late, leaving early or enjoying extended or multiplied breaks during the day can result in many lost hours of productivity per month.

All those points can result in substantial financial loses for companies, and that is where biometrics can help in curbing such a disastrous curve.

Buddy Clocking and Ghost employees, however, can not only be a financial concern but can also result in more severe Occupational Health and Safety issues.
What would happen if somebody rent his job to a non-qualified worker who is supposed to manipulate explosives? Alternatively, if somebody is not fit for the job? If an "acting security guard" has a criminal record?
Such practices could have dire consequences, and the company would ultimately be responsible.
Those issues are quite simple to spot in a small organisation but become a nightmare to manage and eliminate in large organisations of thousands of employees.
With a successful implementation of biometrics, those issues disappear, and companies quickly recover lost money and lost productivity and investments become evident as their ROI is quick.
Let's see some example.

Time & Attendance ROI CALCULATOR

PARAMETERS

Number of employees	1000
Average hours/per day/employee	9
Worked day per month	22
Late arrival in the morning (min)	15
Percentage of late arrival	20%
Leaving early in the evening (minutes)	10
Percentage of early departure	10%
Smoke break time (in minutes)	5
Number of smoking break time/ day	8
Percentage of smoker	10%

PAYROLL COST

Average cost per hour / employee	$ 8	$ 15	$ 25	$ 50
Normal working hours / day	9	9	9	9
Total cost per employee/day	$ 72	$ 135	$ 225	$ 450
Total number of hours per day	9 000	9 000	9 000	9 000
Total cost per day	$ 72 000	$ 135 000	$ 225 000	$ 450 000
Total cost per month	$ 1 584 000	$ 2 970 000	$ 4 950 000	$ 9 900 000
Total cost per year	**$ 19 008 000**	**$ 35 640 000**	**$ 59 400 000**	**$ 118 800 000**

LOST PRODUCTIVITY

Late in the morning (in hours)	50	50	50	50
Cost of late arrival	$ 400.00	$ 750.00	$ 1 250.00	$ 2 500.00
Leaving early in the evening (in hours)	17	17	17	17
Cost of early departure	$ 136	$ 255	$ 425	$ 850
Smoke break (in hours)	67	67	67	67
Smoke Break cost per day	$ 536	$ 1 005	$ 1 675	$ 3 350
Total loss per day (in hours)	134	134	134	134
Total loss per day (in value)	$ 1 072	$ 2 010	$ 3 350	$ 6 700
Total loss per month (in hours)	2 948	2 948	2 948	2 948
Total loss per month (in value)	$ 23 584	$ 44 220	$ 73 700	$ 147 400
Total loss per year (in hours)	35 376	35 376	35 376	35 376
Total lose per year (in value)	$ 283 008	$ 530 640	$ 884 400	$ 1 768 800

Job Costing

Another feature of Time and Attendance that biometrics can help with is Job Costing.
Job Costing refers to the addition of job metrics capture.
Let's take the example of grape harvesting. Employees might be paid a flat rate per day, supplied with food and accommodation and be paid an extra fee based on the volume of grapes they harvest.
In this case, how does one ensure that the right volume is allocated to the right employee?

The process can be integrated into the time and attendance application as follows:

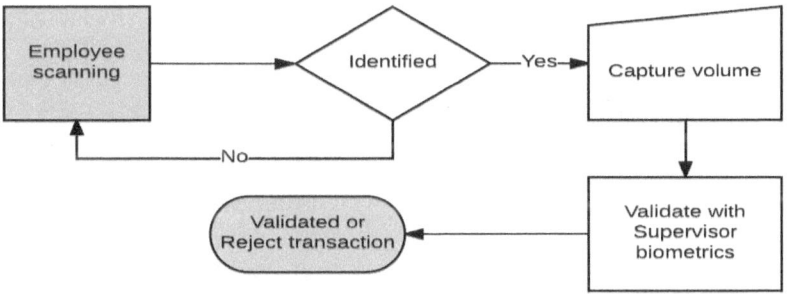

The process described here is simple and more steps could easily be added such as selecting the products that are registered when there are more than one harvested during the same day, i.e. white grape or red grape.
The generated event log should be different from the one covered under the logs sections but had the different information related to the job costing feature (volume, supervisor ID, etc.)
The time and attendance package should be able to process the different transaction from the time of arrival to the time of departure.
It usually happens in the following ways:

By using the first clock of the shift as the clock in and the last one as the clock out.
By manually selecting the "Clock In" option on the reader before clocking in and the "Clock out" option before clocking out.

Scheduling and Resource Planning

Scheduling and Resource Planning can significantly improve reactivity and increase productivity especially on sites where activity can vary at different time of the day.
For example, at ports, all employees clock in at a main location which indicates that they are on site and ready to perform their duties. When a ship docks, available resources can be rapidly redeployed to off-load this specific ship cargo.
Another example could be at the airport, upon arrival of a plane.

Basic Time and Attendance

Once recovered from the reader, the logs are usually saved in a CSV file format.
As its name indicate, a CSV file is a text file filled with values which are separated by commas (Comma Separated Value). This file can be open with a basic text editor such as notepad in Windows or with a worksheet processor such as Microsoft Excel®.
Upon the opening of the file in a worksheet processor, values are displayed in different boxes, and the commas are not displayed.
The main advantage in opening these files in a worksheet processor is that one can manipulate the value to make them more presentable.
Based on the event logs recovered from the reader, one can, therefore, generate a simple report and apply some theme, layout or pivot tables to manipulate and display the data as required.

Advanced Time and Attendance Package

Basic Time and Attendance can be quite limited, and when advanced features like time calculation and exception management are required, it is recommended to make use of an advanced Time and Attendance Package.
In a T&A package, all parameters about local regulation and end-user preferences are adjustable.

Here is a list of parameters that can be customised.

Other timekeeping methods

There are several ways companies track employee time using time tracking software.

- Durational

- Employees enter the duration of the task but not the times when of completion.

- Chronological

- Employees enter start and end times for the task.

Automatic

The system automatically calculates time spent on tasks or whole projects, using a connected device or a personal computer, and user input using start and stop buttons. Users can retrieve logged tasks and view the duration, or the start and stop times.

Exception-based

The system automatically records standard working hours except for approved time off or LOA.

Clock-in clock-out

Employees manually record arrival and departure times.

Monitoring

The system records active and idle time of employees. It might also record screen captures.

Location-based

The system determines the working status of employees based on their location.
Resource-scheduling: by scheduling resources in advance, employees' schedules can be easily converted to timesheets.

Payroll Automation

Payroll refers to all the financial and legal rules and regulations (such as overtime) that a company must follow in order to calculate the amount of money to be paid to each employee at regular interval and after all legal deductions (taxes, contributions, garnish, etc.) have been withheld.
Payroll Automation is used to process all those rules automatically.
Payroll Automation works hand in hand with the Time and Attendance system as it relies on the calculation results given to calculate what is owed accurately.
Today, Payroll automation have become very performant and can prepare, send and print each payslip and if it is linked to internet banking process payment automatically.
When set-up correctly, payroll automation can save time, supply accurate and well-detailed information, eliminate errors and eventually contribute to cost reduction in HR administration cost.
Payroll systems depend on countries/regions but here is a list of some Payroll systems:

- SAP
- Xero
- Sage Payroll
- VIP Payroll
- And other regional and local players

7. Logical Access Control / Identity Management

Logical access control is to the IT world what Physical Access Control is to the real world.
In simpler words, it is all the tools and protocols that allows accessing the resources of an IT system (folders, files, etc.) through identification, verification, authentication, authorisation and other accountability processes. Logical Access Control can be applied globally to a device such as a PC, tablet or other terminal or specifically to a resource or process based on such devices.

(PC) Login

PC Login is a global logical access control because it is at that level that access to a device is given to an individual. At that stage, full access is given to the device, unless restrictions are specifically enforced.

Password / Pin code / Passphrase

Traditionally, to log into a PC, a password is used combined with a username.
The password can be replaced by a pin code or a passphrase depending on the level of security versus the level of convenience that one requires.
The advantage is that username and password are widely accepted and relatively easy to use.
Disadvantages overtake advantages however as passwords can be forgotten quickly, communicated, cracked or even worse, written down and left on the machine so that the user does not forget it but also exposed to every eye.
A password is also referred to as "What you know."

Physical Token based login

Physical tokens such as smart cards (or "What you own") were introduced to improve the level of security. Although those tokens can be used independently to a password, they are usually both used together in conjunction.
This token can also be used to store security components such as certificates without which some resources cannot be accessed.

Biometric-based login

Nowadays, it is not rare to find a biometric reader embedded on our electronic devices. Apple iPhone, Samsung and other brands embed a fingerprint reader into their cell phone, and users mostly embrace it for its security and convenience.
Microsoft launched its solution called "Windows Hello", the goal being is to allow logging in Windows through a certified biometric technology, should it be a fingerprint reader or a camera for facial or iris recognition. Advantages are obvious, but one must be wary of selecting the biometric technology according to its needs as not all biometrics are made equal.

Access to resources

While PC Login gives general access to a device, specific resources such as folders, files or even peripherals can be secured using a password or some biometrics. At the time of access to the resource, the system can request the user to present one or several credentials.

Single Sign-On (SSO)

The Single Sign-on purpose is to allow a user to sign in once to access the single sign-on application and let the application take care of the logging onto software, websites or other services.

It is now widely used by companies like Google with Google Chrome which gives the option to remember usernames and passwords and fill them up during the visit of a website where they are needed.

The clear advantage is convenience and speed of access to different services and data, but the counterpart is that if the primary mean of authentication is hacked, it exposes all the other username and password managed by the SSO application.

Audit process

Logical Access Control and more particularly biometrics can be used for audit purposes.

The main role of an audit is to gather evidence that applicable rules were correctly followed. An audit can, on the other hand, conclude that rules were not followed. An audit must be based on concrete evidence to be undisputable and trusted.

The finding of the audit is summarised in what is called an audit trail.

An audit trail should be comprised of a list of each measurable action that was followed from the beginning of a process to the end.

The problem is to prove that somebody did or did not follow a procedure beyond reasonable doubts.

A couple of examples where an audit trail makes perfect sense.

Firearm issuance.

Law enforcement agencies need access to firearms to carry their duties; however, due to the sensitive nature of firearms, they must keep careful control over the firearms issued.

They need to know whom they issue the firearm to, when and for how long. They also need to know who issued the firearm so that they can ensure authorisation and accountability of all stakeholders in the process.
The system should be able to enforce control points at strategic levels to create a consistent audit trail:

- At computer login point to ensure the right person utilises the computer

- At software startup to ensure the right person logged into the firearm issuance system

- At issuance of the firearm when the issuer confirms the operation as being legit

- At the reception of the firearm to confirm which agent received the firearm

The process goes the over way around when the firearm returns.
This process is all good and well, but without the use of biometrics technology, it becomes a futile exercise.
Indeed, any stakeholder can claim that somebody used his card, pin code, password or any other token instead of him. However, with biometrics, it is not possible.

Note that in this case, the system should be linked to a proper access-controlled locker system to ensure the lockers are only opening when all the necessary checks are confirmed.

Document issuance

Documents (such as credit cards) delivered to a convenient place such as a home or office have become very usual nowadays.
However, to ensure delivery of the document to the right person, biometrics should be included in the tracking solution. Upon delivery, the delivery agent should ask the recipient to authenticate via biometrics.
The biometric feature is enrolled at the time of opening the account, checked at delivery point and a log kept proving that the transaction happened the way it should have.
Without biometrics, it can only be assumed that the document was delivered to the right person.

Transaction authorisation

In any industry where transactions need to be authorised, traditional ways of doing with password or cards can be challenged. In case of a problem, the incriminated person can always claim that somebody used her card or pin-code without her consents. Biometrics is the solution to this problem.
However, not all transactions require the same level of security and this level can be adjusted as per the examples in the table below:

Bank Account Balance	Pin Code
Bank Transfer between account	Pin Code + Password
Bank Payment to another account	Biometrics + Password

Did you Know?

Voice Biometrics allows companies to authenticate a customer transparently while talking on the phone.
Customers pre-enrol and while explaining their requests, the system checks if the recorded voices and the voice on the phone match. If yes, the voice authentication replaces the pin code, and necessary information can be given.

Biometric technologies in healthcare.

In healthcare more than in any other domains it is paramount to ensure the identity of all stakeholders, be it a patient, a nurse, a doctor, etc.
Of course, traditional fingerprint technologies have an essential role to play in securing transactions between patients and practitioners, but when it comes to hygiene concerns, the logical trend is to go contactless.
A few technologies are available to assist in that regards:

- Facial recognition
- Iris recognition
- Contactless fingerprint reader

All these technologies have their pros and cons, and one can learn more about them in the section: "To go further."

Biometric technologies in financial institutions.

In financial institutions, biometrics can be of use for the following application:

- eKYC (Know your customer) during account and new service opening
- Banking applications Staff login
- Transaction authorisation by staff members
- Banking applications Customer login and Transactions authorisation
- Access Control (logical or physical)
- Time and Attendance for staff
- Etc.

Biometric technologies in education institutions.

- Physical or Logical Access Control (staff and students)

- Time and Attendance (staff and students)
- Examination rooms access
- Student identity verification before writing an exam.

Government applications

Civil applications

Civil Application refers to the application used for managing countries populations identity for:

- ID Documents issuance (ID Cards, Passport, etc.)
- Issuing civil documents (driving license, etc.)
- Border control (immigration)
- Voter registration
- Benefits issuance (pensions, welfare etc)

In general, all systems required by the Department of Home Affairs, Foreign Affairs, Road Management, and so forth.

Forensic applications

Forensic applications refer to any police system required to address criminal justice requirement such as:

- Crime scene analysis (to collect and match latent fingerprint)
- Criminal record check (for vetting process purposes)
- Criminal registration

Other Applications

A gateway between commercial application and Government solutions

Some Governments worldwide has to some degrees opened-up access to their citizen databases.
They usually negotiate a concession with private companies who in turn can commercialise the connection as a service. Some Governments prefer to manage this service. However, it is often limited to confirming the validity of an identity.

The two major applications here are:

- Confirmation of someone's Identity

An identity number and a fingerprint template are sent to the government database directly or through the authorised company. The system checks through the biometric feature and sends back a MATCH or no MATCH answer. This service is usually used by organisations such as banks to verify the identity of their customers as part of their eKYC[19] procedure.

- Confirmation of someone's criminal records

Similarly, to confirm someone's identity, companies are specialising in Criminal Records' checks. In this case, the criminal record search is not in a forensics context but for civil purpose. It can be useful during recruitment processes for vacancies requiring screening of candidates.

Both applications should require the consent of the individual to check their identity or criminal records.
It is, therefore, good practice to get the individual to sign an application form or any relevant form of paperwork for future reference.

---

19 Electronically Know Your Customers

Software Integration

In this book, we have covered a few applications using biometrics. However, it was only the tip of the iceberg, and many other applications are possible. Even in applications, we have covered, there are many ways to customise them for specific end-users needs. The customisation is possible via software integration. Manufacturers of biometrics solutions usually propose a Software Development Kit (SDK) to allow system integrators to go further in their developments and integrations.

7. PRIVACY AND DATA PROTECTION

Privacy and data protection have become a significant part of public concern worldwide, and local and regional legislations are quickly evolving to deal with the topic.

In a World where data exchange and communication have exploded over the past decades, people have become extensively anxious to know what is happening to their personal information and who has access to it. The use of biometrics can assist with controlling access to such information, but as with any system, governments are creating legislation and a framework to protect the general public.

There is no shortcut for businesses and compliance is not only ethical but essential to stay out of trouble for a breach can easily destroy a company. The topic is vast, and this section on its own would not be enough to deal comprehensively on the topic.

A whole book would probably be necessary. We will, however, briefly cover some of the existing legislation and organisations in charge of ensuring the enforcement of those rules.

GDPR *(General Data Protection Regulation)*

The General Data Protection Regulation 2016/679 is a regulation in European Union law on data protection and privacy for all individuals within the European Union and the European Economic Area.
This legislation applies to all companies whether they are located within or outside the border of the European Union, as soon as they are dealing with a European Citizen.
This legislation is effective from the 25th of May 2018 and for those companies tempted to think that this does not apply to them, the sanctions planned by the text are:

Administrative fines for non-compliance – tier one

- 2% annual global (Worldwide) turnover or up to 10 million €, whichever is higher.

Administrative fines for non-compliance – tier two

- 4% annual global (Worldwide) turnover or up to 20 million €, whichever is higher.

Fines and enforcement are progressive, meaning that a 20 million € fine might not materialise for a first or minor infringement, but it is worth paying attention.

CNIL *(Commission Nationale Informatique & Liberté)*

The "Commission Nationale de l'informatique et des libertés" is an independent French administrative, regulatory body whose mission is to ensure that data privacy law applies to the collection, storage, and use of personal data. It was founded in France on January the 6th 1978 and is one of the most vigilant and stringent privacy organisation worldwide.

Freedom of Information Act (USA)

The Freedom of Information Act (FOIA) generally provides that any person has the right to request access to federal agency records or information except to the extent the records are protected from disclosure by any of nine exemptions contained in the law or by one of three special law enforcement record exclusions.

Freedom of Information Act (UK)

The main principle behind freedom of information legislation is that people have a right to know about the activities of public authorities unless there is a good reason for them not to. This principle is sometimes described as a presumption or assumption in favour of disclosure.

Right to Information Act (India)

Right to Information (RTI) is an Act of the Parliament of India to provide for setting out the practical regime of the right to information for citizens and replaces the erstwhile Freedom of information Act, 2002. This law was passed by Parliament on 15 June 2005 and came fully into force on 12 October 2005.

POPI Act (South Africa)

The Protection of Personal Information Act is commonly referred to as POPI. The purpose of this Act is to ensure South African businesses responsibly conduct themselves when collecting, processing, storing and sharing another entity's personal information by holding them accountable should they abuse or compromise the third party's personal information in any way.
POPI legislation considers personal information to be 'precious goods' and therefore aims to the owner the personal information, certain rights of protection and the ability to exercise control over the following:

- When and how information is shared (with consent).

- The type and extent of information to be shared (information must only be collected for valid reasons).

- Transparency and accountability on data usage (limited to the purpose it was collected for).

- Providing the owner access to his information (including the right to have his data removed and destroyed).

- Who has access to the information (there must be adequate measures and controls in place to track access and prevent unauthorised people, even within the same company, from accessing the information).

- How and where information is stored

- The integrity and continued accuracy of information (correct capture and update when needed).

Personal information may include but is not limited to:

- Identity and passport number
- date of birth
- phone numbers and email address;
- physical & postal addresses;
- gender, race and ethnic origin;
- criminal record;
- religious or philosophical beliefs including personal and political opinions;
- employment information;
- financial information;
- educational information;
- physical and mental health information including medical history;
- blood type;

As such, the Act defines a 'unique identifier' to be data that "uniquely identifies that data subject in relation to that responsible party." This means that a phone number on its own would not necessarily be personal information, but in combination with a name, for instance, it could be damaging.

It is important to note that the right to protection of personal information is not only applicable to a natural person (i.e. an individual) but any legal entity, including companies and communities or other legally recognised organisations. All of these entities are considered to be 'data subjects' and afforded the same right to protection of their information.

8. FREQUENTLY ASKED QUESTIONS

Is biometric 100% accurate?

Strictly speaking, biometrics is not 100% accurate.
Practically speaking, it can be as accurate as 99.99999999%+
As there are so many factors that can affect performance and accuracy, it is essential to reference major independent benchmarks when selecting a technology. Benchmarks, such as those performed by NIST use large, common datasets to compare a variety of performance aspects across multiple vendors. This is an excellent free resource that can be leveraged when needing to select a biometric technology or vendor.
Refer to section "3. ESSENTIAL CONCEPTS AND LIMITATIONS OF BIOMETRICS" for further information.

Can the error rate be adjusted and measured?

Yes, Error rates are measured through statistics as:

- False Acceptance Rate (FAR) or the possibility for someone to be recognised as somebody else enrolled in the system.
- False Rejection Rate (FRR) or the possibility for someone to be rejected by the system although this individual is enrolled.

Most systems allow configuration of the accuracy of authentication against a trade-off between Security and Convenience.
The higher the security, the less convenient the system and the higher the convenience, the lower the security.
In other words, the higher the FAR, the lower the FRR and vice versa.
When FAR and FRR are equal, it is referred to as Equal Error Rate (ERR).

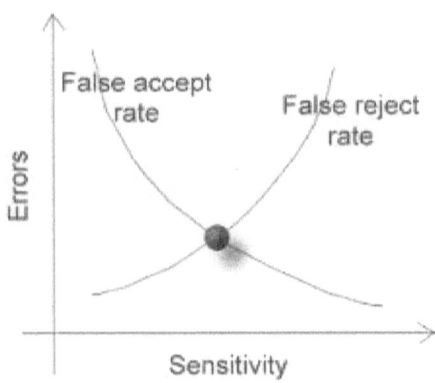

Can everybody be enrolled?

Unfortunately, not everyone can be enrolled in one specific biometric type. However, everybody can be enrolled in at least one biometric type. The rate of individuals who cannot be enrolled on a specific system is called Failure to Enrol (FTE).

Can a template be reverse engineered?

A template by its nature cannot be reverse engineered or in other words; one cannot rework the original picture from a mathematically generated template. The system uses the original picture to start the process but discard it completely by the time the template is generated.

Why can't everybody be enrolled?

To enrol someone, it is a prerequisite that a biometric feature must be available.
If the individual's finger is chopped off, for instance, it would be physically impossible to enrol this individual.

Similarly, any damaged or injured fingerprint might prove challenging to be enrolled. In rare and extreme cases, it might also be impossible to enrol someone's fingerprint. That is the reason why it is paramount to select the right biometric technology for each project.

Are some categories of individuals more inclined to difficulties at enrolment or subject not to be enrollable at all?

For a system to function, it must be set up with parameters to draw a line between what is acceptable and what is not acceptable in term of quality of enrolment. This parameter is called a "Threshold." This threshold is essential as it has a direct impact on the result of authentication (cf. trade-off between security and convenience in the book).
Therefore, any conditions that can drop the quality of enrolment under the threshold will be enough to fail the enrolment:

- skin conditions (eczema, extremely dry or wet skin, etc.)

- harsh working environment (exposition to chemicals, working with bricks, etc.)

- heavy medical treatment (chemotherapy, etc.)

Can these conditions affect FRR and FAR?

Those conditions can affect False Acceptance and False Rejection Rate when they change the biometric feature so much that the reading falls under trade-off threshold set up between Security and Convenience.

What is a template?

A template is a binary file, created from the extraction of a biometric feature. A template cannot be reversed engineered to obtain the original picture.

What is the difference between a template and an algorithm?

An algorithm is a formula or the recipe that a computer uses to achieve a goal or a result.
In this case, the result is the template.
Very often, people interchange both terms template and algorithm to describe a template.
This is an abuse of language and is incorrect.

Is there any known issue with juveniles?

The main issue with juveniles is that their body is still forming. Hence, it might be difficult sometimes to enrol them if the biometric selected is not developed enough (too small).

Furthermore, the natural development of juveniles' body often results in the reference template no longer being relevant after some time and usually forces the re-enrollment of the individual after a couple of months.

To limit the need for re-enrollment, manufacturers have come up with what is called "juvenile mode" that artificially calculates the expected future template based on the reference template. This helps in minimising the need for re-enrolment but does not suppress it entirely.

Can grease/oil affect the reading of fingerprint Biometric?

Grease/Oil in reasonable dosage should to the contrary enhance the reading of a fingerprint.
In excess, however, it could affect the reading, especially if it blurs or cover the fingerprint itself.

What is the difference between 2D and 3D Facial recognition?

2D facial is based on the analysis of a basic 2D picture whereas 3D facial recognition is based on the analysis of an advanced 3D facial mesh reconstruction.
3D facial is much more secure than 2D and cannot be spoofed as easily.
2D facial can easily be spoofed even with some anti-spoofing mechanism.
However, both find their application mainly due to costs differences.

Can somebody copy fingerprints or any other biometrics?

In theory, biometrics can be copied however some are more difficult to reproduce than others.

For instance, it might be easy to reproduce a 2D picture, but it should not be that easy to reproduce a DNA sequence (or at least not available to the common public and at an affordable cost).

Therefore, common sense shall prevail, and biometric technology should be selected based on each use case.

Manufacturers have developed anti-spoofing mechanism that should be able to differentiate between a genuine or non-genuine biometric feature.

For instance, for fingerprint, there is what is called FFD (Fake Finger Detection). The feature is described in detail in the book.

Can tattoos be used as a biometric feature?

Tattoos are not classified as a primary biometric feature as such as they cannot uniquely identify individuals (two people can wear the same tattoos). They can however sometimes be used to identify people when no other biometric features are available. This could be the case in some criminal investigations where the suspect is not known, and witnesses can describe a tattoo.

9. TO GO FURTHER

Further information about Fingerprint

Fingerprint Based enrolment

With a fingerprint-based system, there are two bits of information which can use for enrolment:

- minutiae points
- pattern based

Minutiae point

Minutiae points are all characteristics or unique points found on the fingerprint and classified as per the following picture:

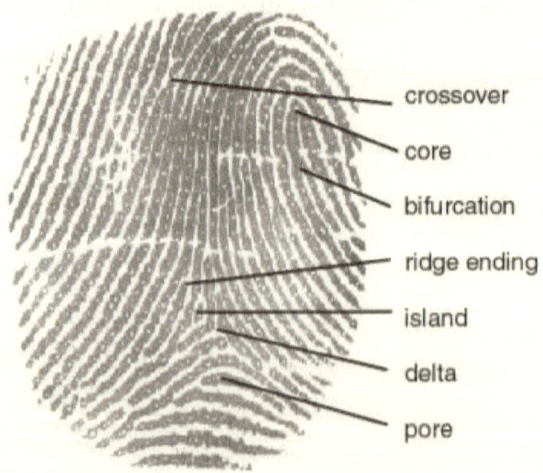

Each point is referenced similarly to coordinates on a grid.

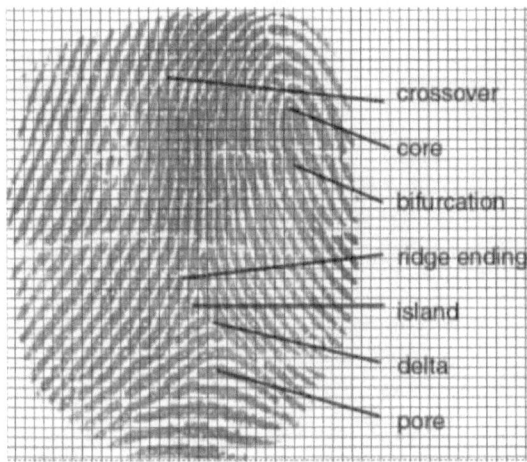

Subsequently, some connecting lines and angles between coordinates are drawn and used to generate a binary file made of 1 and 0.

Patterns

A pattern-based fingerprint system does not rely on specific characteristic points or minutiae points but instead on shapes and directions of ridges on the fingerprint.

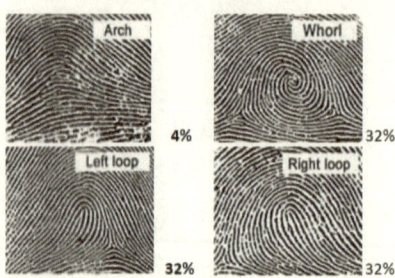

Those four classifications can have sub-categories such as:

- Plain Arch
- Tented Arch
- Ulnar Loop
- Radial Loop
- Plain Whorl
- Central Pocket Loop
- Double Loop Whorl
- Accidental Whorl

Processing false information

As we have seen earlier in this book, biometric systems work based on statistics.

In the case of a fingerprint-based system, it works based on information presented to it. Now, what happens if a finger is excessively dirty or injured. The system may be interpreting some false minutiae points as true as per the picture below. The picture on the left represents the original fingerprint with no injury and 14 minutiae points while the picture on the right represents the same fingerprint with an injury and therefore 28 minutia points. This example illustrates the creation of fake minutia points and the dots have been positioned randomly.

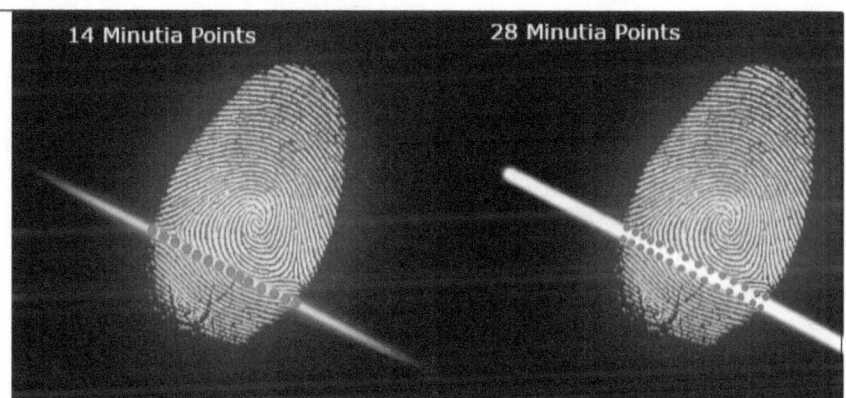

Fingerprint (14 Minutia Points) vs the same Fingerprint with Injuries (28 Minutia Points)

Powerful algorithms can differentiate false minutia points resulted from injuries or dirty fingers from valid minutia points. In this case, a robust algorithm would be able to virtually heal the injury and only count 14 minutiae points instead of 28.

Physical Characteristics

FINGERPRINT			
Technology / Characteristic	Pros	Cons	Comment
Optical	• Easy and Convenient to use • Fairly Cheap • Accepted by everybody • Small template	• Difficulties with dry and wet fingers • Doesn't function if fingerprints are not visible (gloves, no print, etc., fingers too dirty.) • Fingerprints can be affected by dread disease treatment (chemotherapy, etc.)	Although it is not as easy as you can see in movies, fingers can be replicated artificially and then form of what is called a Fake Finger.
Multispectral	• Can read on the subdermal layer of the skin • Suitable for dry and wet finger	• Cannot be FBI PIV IQS • Significantly more expensive than other fingerprint sensor technologies • Can be slower than other fingerprint sensor technologies	Why can MS not be FBI PIV IQS Certified? Because PIV IQS certification requires a minimum image quality which multispectral fails to reach as it technologically doesn't look at one image but collate information from different wavelengths and reconstitute an image. This process does lower the quality of the

				picture.
Capacitive	• Cheapest fingerprint sensor technology. • Lightweight • Low power consumption • Embeddable where size matters (smartcards, etc.)	• Fragile • Sensitive to high Electrostatic Discharge (ESD) • Small scanning area (challenging to scan big fingers accurately)	Not for large-scale systems deployment	
Swipe	• Small • Easy to use	• Cheap • Fragile	Not for large-scale systems deployment	
Contactless	• Very Fast,	• Relatively high cost		

	Pros	Cons	Comment
	high throughput optimised • Hygienic (due to no contact) • Easy to use • Some products are designed to be unaffected by sun and dust • Privacy • Usable at arm's length only • Requires deliberate action • Usually works on many aspects of the fingers (fingerprint, curvature, shape, etc.)	compared to traditional fingerprint technologies • Doesn't function if fingerprint is not visible (gloves, no print, etc., fingers too dirty.) • The fingerprint can be affected by dread disease treatments (chemotherapy, etc.)	

VEIN PRINT			
Technology / Characteristic	Pros	Cons	Comment
Finger / Palm / Wrist	• Very Secured • Very difficult to spoof	• Relatively high cost compared to traditional fingerprint technologies • Veins can be affected by dread disease treatment	

| | | (chemotherapy, etc.) | |

HAND GEOMETRY

Technology / Characteristic	Pros	Cons	Comment
Hand or finger Geometry	• Easy to use • Dry or wet skin is not an issue •	• Not unique so can only be used for verification • Not great for children • Accessories such as rings may prevent usage • Not ideal for use for people with • Large device	Works on the geometry of the hand.

FACIAL

Technology / Characteristic	Pros	Cons	Comment
2D Facial	• Fairly cheap • Easy to use • Easy to integrate (standard camera or webcam suffice)	• Fairly Easy to spoof •	
3D Facial	• Very Secure • Easy to use • Robust performances	• Expensive Technology • Can be bulky	

EYES

Technology / Characteristic	Pros	Cons	Comment
IRIS	• Very secure • 2 x biometrics	• Near-infrared technology so	Some cultures believe that taking

	features (2 eyes)	light control can be an issue • Intrusive • Requires the user proactiveness, which is not always possible • Sometimes culturally unacceptable	a photograph of their face or eyes is like stealing their soul.
RETINA	• Very accurate • Not exposed to external factors (dust, etc.) •	• Very intrusive • Requires proactiveness from the user (this is not always possible) • Sometimes culturally unacceptable	Some cultures believe that taking a photograph of their face or eyes is like stealing their soul.

DNA			
Technology / Characteristic	Pros	Cons	Comment
DNA	• Extremely accurate • Can be used with very small samples	• Long process • Costly	Due to its long process (from several hours to several days), DNA recognition technologies are currently not an option for access control/time and attendance.

Behavioural Characteristics

Technology / Characteristic	Pros	Cons	Comment
VOICE	• Easy and Convenient • Cheap	• Fairly easy to spoof • Can be affected by surrounding noise	
GAIT	• Works remotely • Effective even with low-quality images • Non-invasive	• Accuracy may be low • Gait may be altered by clothes or shoes (high heel vs flat shoes, jeans & T-shirt vs Trousers & Thick Coat)	
KEYSTROKE	• Free (Keyboard typing) • Easily integrable in an IT solution	• Low accuracy • Keystrokes can change based on physiological factors (fatigue, etc.) • The dynamic may change on different keyboards (QWERTY, AZERTY, etc.)	
SIGNATURE	• Inexpensive / free • Widely accepted as legal	• Unsecured and easily imitable • People may not always sign exactly, in the same way, depending on external factors (orientation of the signature space, utensils used, etc.)	

Biometrics Comparison Table

Biometrics	Accuracy	Cost	Acceptability	Permanence
Fingerprint	High	$$	High	High
Contactless FP	High	$$$$	High	High
Vein	High	$$$	Medium	High
Facial 2D	Low	$	High	Medium
Facial 3D	Medium	$$$$$	High	Medium
Iris	High	$$$$	Medium-Low	High
Retina	High	$$$$	Low	High
DNA	Extremely High	$$$$	Low	High
Voice	Medium	$	High	Medium
Gait	Low	$	High	Medium
Hand key	Medium - low	$$	High	Medium
Signature	Low	$	High	Medium

10. CONCLUSION

What does the future of Biometrics hold?
That is the very big question to which one can only try to give some answers based on current trends in biometrics technologies but also in other technologies such as Information Technologies, Artificial Intelligence, and so forth.

The first trend is Contactless technologies. Due to hygiene concerns, and the expectation for increased convenience, contactless technologies are more and more in demand. Their higher cost than traditional contact biometrics has delayed mass adoption but as technologies mature and prices come down, so the adoption rate increases.

The second trend is the extension of some technologies such as video cameras to capture biometric integration through various applications. Camera's manufacturers are adding the capabilities to their device to recognise an individual through the regular video feed. It is not perfect yet and still needs to be used in a controlled environment (lighting conditions, camera orientation, etc.), but it will undoubtedly improve in the future.

The third trend is the miniaturisation of biometric technologies. Miniaturisation is not new and has been happening for years, but it has now reached a breaking point where it has become viable to use biometrics in devices where it once was considered as a gadget.
Biometrics is nowadays present everywhere, and its miniaturisation has allowed for integration into cellular phones and other portable devices. Cell phone manufacturers are now looking at integrating the biometric imager within the glass of the device. The camera of a cell phone can already be used as an imager by capturing fingerprint, face or iris.

Smart card manufacturers are now proposing embedded fingerprint readers on payment cards. The card can be used by placing it in the payment terminal and holding the card between the index and the thumb, placing the thumb on the fingerprint reader for reading. The fingerprint replaces the pin code associated to the card. Manufacturers have pushed even further, and with compatible terminals, one can tap a card (wireless transaction) while still enjoying a secure transaction with simultaneous biometric authentication.

The fourth trend is the increasing demand for BYOD[20] - "Bring Your Own Device". Companies are moving slowly in allowing their employees to use their own devices, particularly due to the rate at which modern consumers upgrade their cell phones, and the associated administration and security burden. Managing such a diverse and evolving set of devices, particularly if biometric data is stored therein, introduces a significant number of variables which are not present when standardising on a specific reader or token technology.

Whichever direction the technology takes, biometrics are here to stay, and they are going to become increasingly part of our lives. We have never been in a more exciting era of technology.
Biometrics and Access Control are more than aligned with IT systems and the advent of technologies such as Artificial Intelligence, deep learning, machine learning, blockchain and artificial neural network, will also no doubt have a real impact on performances and accuracy not only of the authentication process but also of decision-making capabilities in applications such as BSIM[21].

At the same time, it is important to remember that Biometrics is but one extra tool available to secure a system or a device.
There are always advantages and shortfalls in any system or components of this system, and ultimately the responsibility lies on the customer's shoulders to choose what is best adapted to suit their needs. However, how can a someone who is not specialised in this domain be able to make the right decision without assistance?

---

20 - BYOD : Bring Your Own Device
21 - BSIM : Business & Security Identity Management

I see here a great opportunity for market players, through genuine expert advice and solutions customisation to create long-term relationships with those customers. This is true at every level of the chain: manufacturers, distributors, system integrators, installers, consultants, etc.

TRUST and ADDED-VALUE are the key words here.
Happy customers are always long-term customers.

I hope you have enjoyed reading this book and that it gave a good understanding of what biometrics is all about and how it links to real life application such as Access Control or Time & Attendance.

11. REFERENCES & CREDITS

USA Standards:

https://www.fbibiospecs.cjis.gov/Certifications/FAQ

German Standards

https://www.bsi.bund.de/EN/Publications/TechnicalGuidelines/TR03121/BSITR03121.html

Indian Standards

http://egovstandards.gov.in/biometrics

EMV (Europay, Mastercard, VISA

https://www.emvco.com/

Images used in this book

Images used on this website are either licensed under the common creative license and of unknown author or images dully licensed for which the author has acquired the usage rights.

https://creativecommons.org/licenses/by-sa/4.0

Other Websites

https://www.i-learn-online.com

http://www.findbiometrics.com

12. DEFINITIONS

Access Control:

Restriction of access to physical or logical resources based on a set of rules, usually following questions such as:

- Who?
- What?
- When?
- Where?
- How?

Algorithm:

An algorithm is to the high-tech world what a recipe is to the cooking world. In other words, an algorithm is a set of instructions, described step by step and which goal is to explain to a computer how to reach the desired result, i.e. how to find a fingerprint amongst a database of fingerprints.

Biometrics:

The word "Biometrics" is composed of two particles described as follows:

- "Bio," short of "Biological" or "Life" and by extension "Body."
- "Metric" for "Measurement"

Biometrics stands for the "Measurement of body characteristics."
In our modern world, Mr Bertillon (French) was the first to use biometric for the measurement of body parts (arms, legs, skull perimeter, etc.).
Mr Galton (British) by extension worked on the comparison of fingerprints.
Body characteristics are much more difficult to give, sell, steal, copy, reproduce, etc. than any other mean of identification.

Bit

A bit is the smallest unit of data in a computer. It has a binary value of 1 or 0 (ON or OFF).
In telecommunication, the bit rate is the number of bits which are transmitted within a specific time frame and this is usually measured in bits per seconds. The higher the rate, the faster the connection.
We often refer to systems as 8bits, 16bits, 32bits, 64bits and sometimes 128bits. The Operating System of the device must, therefore, be developed to be compatible with the chosen bit versions (Windows 32-bits, Windows 64-bits, Linux 64-bits, etc.)
In binary, 1 Kbit is not 1000 bit. It is actually, 2^{10} bit or 1024 bit.

1024 Bit = 1 Kib (Kibibit)   1024 Kib = 1 Mib (Mebibit)

1024 Mib = 1 Gib (Gibibit)   1024 Gib = 1 Tib (Tebibit)

1024 Tib = 1 Pib (Pebibit)   1024 Pib = 1 Eib (Exbibit)

Byte

A byte is a unit of digital information storage made of 8 bits. The Byte unit is commonly used to quantify the size of a file or any storage media. For instance, a 2Tbyte Hard Drive.
As for the bit, 1Kbyte is not equal to 1000 byte but to 1024 bytes as per the table below.

1024 Bytes = 1 KB (Kilobyte)   1024 KB = 1 MB (Megabyte)

1024 MB = 1 GB (Gigabyte)   1024 GB = 1 TB (Terabyte)

1024 TB = 1 PB (Petabyte)   1024 PB = 1 EB (Exabyte)

1024 EB = 1 ZB (Zettabyte)   1024 ZB = 1 YB (Yottabyte)

BRING YOUR OWN DEVICE (BYOD)

BYOD refers to a policy which allows employees to bring their personal devices to their workplace and to use those devices to access company resources (software, data, etc.)
Security is an issue that companies should take seriously and therefore use of advanced security systems such as biometrics, encryption, and remote control of the devices (possibility to wipe their content out if lost or stolen) is frequently used by company embracing BYOD policy.

CLASSIFICATION:

Fingerprints are classified into different shapes.
Fingerprints divide into four main categories.

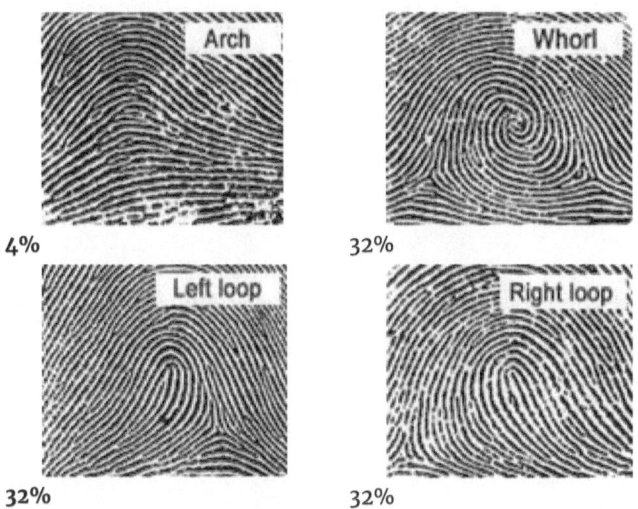

Those four classifications can have sub-categories such as:

- Plain Arch
- Tented Arch
- Ulnar Loop
- Radial Loop
- Plain Whorl
- Central Pocket Loop
- Double Loop Whorl
- Accidental Whorl

COMMISSIONING

Commissioning is the phase of a project where the project implementation is checked and where usually an independent body (auditor or manufacturer or other) confirm that the workmanship has been completed according to the industry standard to ensure that the system runs according to specifications before officially declaring the end of the project.

DECOMMISSIONING

Decommissioning is the action of uninstalling a site and leave it in the same condition as it was before the installation of a system.

DUPLEX (HALF/FULL)

Full Duplex is a mode of communication which allows to simultaneously send and receive messages on separate lines (think of a phone conversation where one can talk and listen at the same time).
Half Duplex is only allowing to send or receive at one given time and does not allow simultaneous communication (think of a walkie-talkie communication where you talk then listen).

ENROLMENT:

Enrolment is the phase in which biometric features are acquired from an individual and converted into a digital format. This digital format is a binary file called "template" and which is composed of "0" and "1," e.g., "1010110011". The template is then stored in a database for future reference. The size of the template will depend on the type of biometrics captured as well as the level of compression of the file.

EER (Equal Error Rate):

Equal Error Rate is, in theory, the rate where a biometric feature has as much chance as being rejected than accepted. On a scale from 1 to 10, it would be at 5.

FAR (False Acceptance Rate):
False Acceptance Rate is, in theory, the rate at which somebody is presenting his/her biometric feature will be identified/verified while not supposed to.
I insist here on the word "in theory" because it means that if you place your finger 100 times, x% of the time, you will be accepted where you are not in the database.
However, this percentage is very much dependent on many more parameters like the quality of the templates, how many templates are the database, the environment, etc.
In an efficient system, it is possible to tune the level of FAR from 1% to 1 per million, and it is spread over a scale going from 1 to 10.

Firmware

A Firmware is the operating system of a device. It is the software that allows the device to function within its specifications.

FFD (Fake Finger Detection):

Fake Finger Detection is a feature that allows for the detection of fake fingers or attempted.
There are three types of Fake finger detection. See spoofing section for more information.
FRR (False Rejection Rate):

False Rejection Rate is, in theory, the rate at which somebody is presenting his/her biometric feature will be rejected while not supposed to. FRR cannot be set and is a direct consequence of the FAR rate.

FTE (Failure to Enrol):

Any biometric system, depending on the population being enrolled, will always have a percentage of people that cannot be enrolled.
This rate is referred to as **FTE** or Failure to Enrol.
As an example, the fingerprints of people working with chemicals and their fingerprint are worn which makes it difficult, if not impossible to enrol fingerprint; the result is FTE (finger vein, etc.).
Most reputable manufacturers will, however, propose alternatives to counter those difficulties.
Extreme cases may arise, and if unfortunately, an individual does not have hands, therefore there would be no possibility to enrol a finger. Even in this case, the individual can use alternative biometric features such as facial or iris recognition. The manufacturer or the reseller should be able to assist with the best-suited solution for each use case.

IDENTIFICATION

Can also be pronounced 1 to N or 1 to many and written 1:n or 1:m.
Identification means looking for a record amongst many records.
No token is presented to operate, and the system must find the individual credential or declare that the individual is not part of the database ("No Hit").

From a theoretical point of view, **identification** is a succession of verifications.
In this case, verification will be performed on each record of the database one after the other until the person is identified or until the end of the database, in which case the system will return a "NOHIT" message.

OPTRONIC LAYER

An optronic layer is an extra layer added onto an optical sensor.
This layer has the appearance of an electronic circuit and is used in Hardware Fake Finger Detection system to detect if a finger is real or not. It is typically able to measure the impedance of the finger, etc.
A piece of plastic or rubber does not have impedance and in this case cannot be confounded with a real finger.

OSDP (Open Supervised Device Protocol)

Open Supervised Device Protocol (OSDP) is an access control communications standard developed by the Security Industry Association (SIA) to improve interoperability among access control and security products. OSDP v2.1.7 is currently in process to become a standard recognised by the American National Standards Institute (ANSI), and OSDP is in constant refinement to retain its industry-leading position.

ONVIF® – Profile A

A fixed set of functionalities describes an ONVIF profile through numerous services that are provided by the ONVIF standard. Several services and functionalities are mandatory for each type of ONVIF profile. An ONVIF device and client may support any combination of profiles and other optional services and functionalities.
Profile-A is the ONVIF Profile dedicated to Access Control.

POPULATION

We call population a group of people susceptible to enrol and use a system.
A population can be:

- a group of employees in a company
- a group of students at a University
- a country population
- and so forth

TAILGATING

Tailgating or piggybacking is the passage of unauthorised users behind a genuinely authorised user.
This is one of the most widespread security breaches affecting businesses today.
Turnstiles and infrared detection systems are two ways of limiting tailgating.

TEMPLATE

A template is a binary file which represents a biometric feature.
This binary file is called "template" and is composed of "0" and "1," e.g., "1010110011".

VERIFICATION

Verification is also pronounced 1 to 1 and written 1:1
This is the action of verifying someone against one record stored in a token. The token can be an ID card or any other documents that can be read either by a human or by a machine. There is, therefore, a verification between one person who claims to be someone and the information of the person stored on the token.

13. ABOUT THE AUTHOR

Nicolas started his career in France as an Information Technology Specialist when floppy disks were still in use. In 2002, he moved to South Africa where he worked for the French Embassy – Trade Commission as IT Manager for Southern Africa operations for two years.

Nicolas was subsequently introduced to Access Control and Biometrics in 2004 when he was appointed Technical Manager by a French company, SAGEM South Africa, a global leader in Biometrics when they were trying to enter the African market.

Nicolas draws his legitimacy as an expert in his field from his involvement in Information Technology, Biometrics, Access Control and Time and Attendance. Over the past two decades, he has occupied key positions such as Technical Manager, Sales and Operation Manager, Sales and Marketing Director, Business Unit Director, and Regional Director of Sales in Africa at various companies leaders in their market.

Nicolas is well known and respected within the industry and regularly contacted by consultants, end-users, and other professionals for his knowledge and expertise in the field of biometric and security technologies. Nicolas has been travelling extensively all other the world to promote biometrics and more broadly in Africa including but not limited to South Africa, Botswana, Kenya, Tanzania, Namibia, Malawi, Democratic Republic of Congo, Cameroon, Ghana, and Zimbabwe.

www.ingramcontent.com/pod-product-compliance
Lightning Source LLC
Chambersburg PA
CBHW031416210526
45464CB00005B/1902